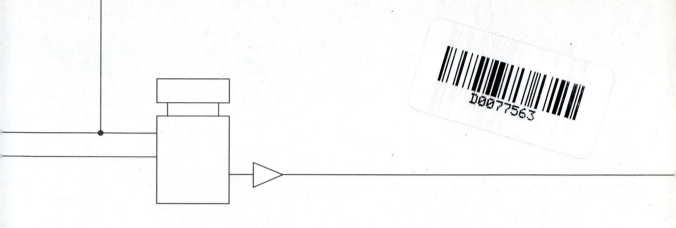

The 8051 Microcontroller
ARCHITECTURE, PROGRAMMING,
and APPLICATIONS

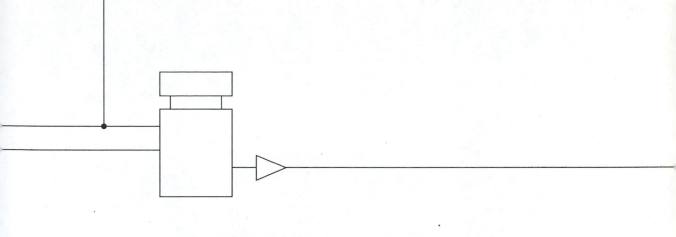

The 8051 Microcontroller
ARCHITECTURE, PROGRAMMING, and APPLICATIONS

Kenneth J. Ayala
Western Carolina University

WEST PUBLISHING COMPANY

ST. PAUL ▪ NEW YORK ▪ LOS ANGELES ▪ SAN FRANCISCO

Copyediting: Technical Texts, Inc.
Text and Cover Design: Roslyn Stendahl, Dapper Design
Cover Image: Christopher Springmann, The Stock Market
Composition: G & S Typesetters, Inc.
Artwork: George Barile, Accurate Art

COPYRIGHT © 1991 By   WEST PUBLISHING COMPANY
610 Opperman Drive
P.O. Box 64526
St. Paul, MN 55164-0526

Printed in the United States of America ∞

98 97 96 95 94 93   8 7 6 5 4 3 2 1

Library of Congress Cataloging-in-Publication Data

Ayala, Kenneth J.
   The 8051 microcontroller : architecture, programming,
and applications / Kenneth J. Ayala.
     p.   cm.
   Includes index.
   ISBN 0-314-77278-2 (soft)
   1. Intel 8051 (Computer) 2. Digital control systems.
I. Title.
QA76.8.I27A93   1991
004.165—dc20   90-12928
                               CIP

To John Jamison of VMI and
John Peatman of Georgia Tech,
both of whom made this book possible

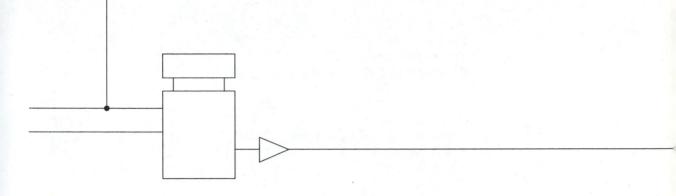

Contents

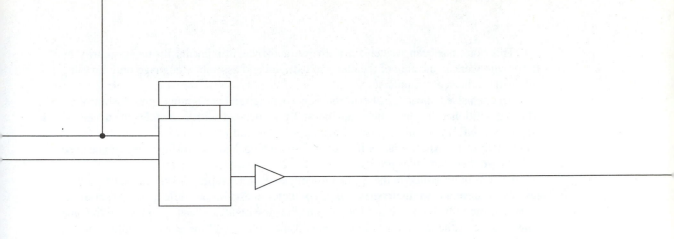

Preface

The microprocessor has been with us for some fifteen years now, growing from an awkward 4-bit child to a robust 32-bit adult. Soon, 64- and then 128-bit wizards will appear to crunch numbers, spreadsheets, and, CAD CAM. The engineering community became aware of, and enamored with, the 8-bit microprocessors of the middle to late 1970's. The bit size, cost, and power of these early CPUs were particularly useful for specific tasks involving data gathering, machine control, human interaction, and many other applications that granted a limited intelligence to machines and appliances.

The personal computer that was spawned by the 8-bit units predictably became faster by increasing data word size and more complex by the addition of operating system hardware. This process evolved complex CPUs that are poorly suited to dedicated applications and more applicable to the generic realm of the computer scientist and system programmer. Engineering applications, however, did not change; these applications continue to be best served by 8-bit CPUs with limited memory size and I/O power. Cost per unit also continues to dominate processing considerations. Using an expensive 32-bit microprocessor to perform functions that can be as efficiently served by an inexpensive 8-bit microcontroller will doom the 32-bit product to failure in any competitive marketplace.

Many designers continue to use the older families of 8-bit microprocessors. The 8085, 6502, 6800, and Z80 are familiar friends to those of us who had our first successes with these radical new computers. We know their faults and idiosyncrasies; we have, quite literally, tons of application software written for them. We are reluctant to abandon this investment in time and money.

New technology makes possible, however, a better type of small computer—one with not only the CPU on the chip, but RAM, ROM, Timers, UARTS, Ports, and other common peripheral I/O functions also. The microprocessor has become the microcontroller.

Some manufacturers, hoping to capitalize on our software investment, have brought out families of microcontrollers that are software compatible with the older microprocessors. Others, wishing to optimize the instruction set and architecture to improve speed and reduce code size, produced totally new designs that had little in common with their earlier microprocessors. Both of these trends continue.

This book has been written for a diverse audience. It is meant for use primarily by those who work in the area of the electronic design and assembly language programming of small, dedicated computers.

An extensive knowledge of electronics is not required to program the microcontroller. Many practitioners in disciplines not normally associated with computer electronics—transportation, HVAC, mechanisms, medicine, and manufacturing processes of all types—can benefit from a knowledge of how these "smart chips" work and how they can be used to improve their particular product.

Persons quite skilled in the application of classical microprocessors, as well as novice users who have a basic understanding of computer operation but little actual experience, should all find this book useful. The seasoned professional can read Chapter 2 with some care, glance at the mnemonics in Chapters 3 through 6, and inspect the applications in Chapters 7, 8, and 9. The student may wish to quickly read Chapter 2, study the mnemonics and program examples carefully in Chapters 3 through 6, and then exercise the example programs in Chapters 7, 8, and 9 to see how it all works.

The text is suitable for a one- or two-semester course in microcontrollers. A two semester course sequence could involve the study of Chapters 1 to 6 in the first semester and Chapters 7, 8, and 9 in the second semester in conjunction with several involved student programs. A one-semester course might stop with Chapter 7 and use many short student assignments drawn from the problems at the end of each chapter. The only prerequisite would be introductory topics concerning the basic organization and operation of any digital computer and a working knowledge of using a PC compatible personal computer.

No matter what the interest level, I hope all groups will enjoy using the software that has been included on a floppy disk as part of the text. It is my belief that one should not have to buy unique hardware evaluation boards, or other hardware-specific items, in order to "try out" a new microcontroller. I also believe that it is important to get to the job of writing code as easily and as quickly as possible. The time spent learning to use the hardware board, board monitor, board communication software, and other boring overhead is time taken from learning to write code for the microcontroller.

The programs included on the disk, an 8051 assembler named A51, and a simulator, named S51, were both written by David Akey of PseudoCorp, Newport News, Virginia. PseudoCorp has provided us all with a software development environment that is not only easy to use but one that we can uniquely configure for our own special purposes. Details on the assembler and simulator are provided in the proper appendixes; use them as early as possible in your studies. Many points that are awkward to explain verbally become clear when you see them work in the simulator windows! Further information on products developed by PseudoCorp follows this Preface.

I have purposefully not included a great deal of hardware-specific information with the text. If your studies include building working systems that interface digital logic to the microcontroller, you will become very aware of the need for precise understanding of the electrical loading and timing requirements of an operating microcontroller. These details are best discussed in the manufacturer's data book(s) for the microcontroller and any associated memories and interface logic. Timing and loading considerations are not trivial; an experienced designer is required to configure a system that will work reliably. Hopefully, many readers will be from outside the area of electronic design and are mainly concerned with the essentials of programming and interfacing a microcontroller. For these users, I would recommend the purchase of complete boards that have the electrical design completed and clear directions as to how to interface common I/O circuits.

Many people have played a part in writing this book. Special thanks go to all of the following people:

The reviewers of the early, really rough, drafts of the text:
Richard Barnett, Purdue University
Richard Castellucis, Southern College of Technology
Jerry Cockrell, Indiana State University
James Grover, University of Akron

Chris Conant, Broome Community College–New York
Alan Cocchetto, Alfred State College–New York
for their thoughtful criticisms and words of encouragement.

Cecil A. Moore, Staff Applications Engineer for Intel Corporation in Chandler, Arizona, whose meticulous comments have greatly improved the technical accuracy and readability of the text.

Tom Tucker of West Publishing for his willingness to experiment.

Anne, my wife, for many years of patience and understanding.

My students, past and present, who have taught me much more than I have taught them.

Finally, let me thank you, the reader. I would be very grateful if any errors of omission or commission are gently pointed out to me by letter or telephone. Thank you for your help.

Kenneth J. Ayala
Western Carolina University
Cullowhee, North Carolina

PseudoCorp Software Development Tools

PseudoSam Cross-Assemblers

Our line of macro cross-assemblers are easy to use and full featured, including conditional assembly and unlimited include files. Available for most processors, these assemblers provide the sophistication and performance expected of a much higher priced package.

PseudoMax Simulators

Don't wait until the hardware is finished to debug your software. Our simulators can test your program logic on your PC before the target hardware is built.

PseudoSid Disassemblers

A minor glitch has shown up in the firmware, and you can't find the original source program. Our line of disassemblers can help you re-create the original assembly language source.

DeveloperPack

Buy our developer package and the next time your boss says "Get to work!", you'll be ready for anything. Each DeveloperPack contains a cross-assembler, simulator, and disassembler for a particular microprocessor.

Broad Range of Support

Intel 8048	RCA 1802,05	Intel 8051	Intel 8096
Motorola 6800	Motorola 6801	Motorola 68HC11	Motorola 6805
Hitachi 6301	Motorola 6809	MOS Tech. 6502	WDC 65C02
Rockwell 65C02	Intel 8080,85	Zilog Z80	NSC 800
Hitachi HD64180	Motorola 68000,8	Motorola 68010	Intel 80196

and more in development.

To Be Released in 1991

Pascal Cross-Compilers
C Cross-Compilers
Advanced Simulators
Linking Assemblers

PseudoCorp Software

To Order Or For More Information:

PseudoCorp
Professional Development Products Group
716 Thimble Shoals Blvd., Suite E
Newport News, VA 23606

Telephone: (804)873-1947   FAX: (804)873-2154

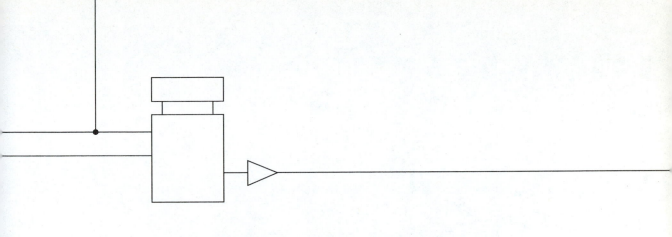

The 8051 Microcontroller
ARCHITECTURE, PROGRAMMING, and APPLICATIONS

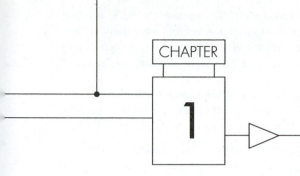

CHAPTER

1

Microprocessors and Microcontrollers

Chapter Outline

Introduction
Microprocessors and Microcontrollers
The Z80 and the 8051
A Microcontroller Survey

Development Systems for Microcontrollers
Summary

Introduction

The past two decades have seen the introduction of a technology that has radically changed the way in which we analyze and control the world around us. Born of parallel developments in computer architecture and integrated circuit fabrication, the microprocessor, or "computer on a chip," first became a commercial reality in 1971 with the introduction of the 4-bit 4004 by a small, unknown company by the name of Intel Corporation. Other, more well-established, semiconductor firms soon followed Intel's pioneering technology so that by the late 1970s one could choose from a half dozen or so microprocessor types.

The 1970s also saw the growth of the number of personal computer users from a handful of hobbyists and "hackers" to millions of business, industrial, governmental, defense, educational, and private users now enjoying the advantages of inexpensive computing.

A by-product of microprocessor development was the microcontroller. The same fabrication techniques and programming concepts that make possible the general-purpose microprocessor also yielded the microcontroller.

Microcontrollers are not as well known to the general public, or even the technical community, as are the more glamorous microprocessors. The public is, however, very well aware that "something" is responsible for all of the smart VCRs, clock radios, washers and dryers, video games, telephones, microwaves, TVs, automobiles, toys, vending machines, copiers, elevators, irons, and a myriad of other articles that have suddenly become intelligent and "programmable." Companies are also aware that being competitive in this age of the microchip requires their products, or the machinery they use to make those products, to have some "smarts."

The purpose of this chapter is to introduce the concept of a microcontroller and survey a representative group. The remainder of the book will study one of the most popular types, the 8051, in detail.

Microprocessors and Microcontrollers

Microprocessors and microcontrollers stem from the same basic idea, are made by the same people, and are sold to the same types of system designers and programmers. What is the difference between the two?

Microprocessors

A *microprocessor,* as the term has come to be known, is a general-purpose digital computer central processing unit (CPU). Although popularly known as a "computer on a chip," the microprocessor is in no sense a complete digital computer.

Figure 1.1 shows a block diagram of a microprocessor CPU, which contains an arithmetic and logic unit (ALU), a program counter (PC), a stack pointer (SP), some working registers, a clock timing circuit, and interrupt circuits.

To make a complete microcomputer, one must add memory, usually read-only program memory (ROM) and random-access data memory (RAM), memory decoders, an oscillator, and a number of input/output (I/O) devices, such as parallel and serial data ports. Additionally, special-purpose devices, such as interrupt handlers and counters, may

FIGURE 1.1   A Block Diagram of a Microprocessor

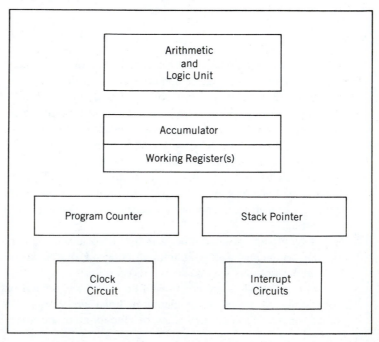

be added to relieve the CPU from time-consuming counting or timing chores. Equipping the microcomputer with a mass storage device, commonly a floppy disk drive, and I/O peripherals, such as a keyboard and a CRT display, yields a small computer that can be applied to a range of general-purpose software applications.

The key term in describing the design of the microprocessor is *"general-purpose."* The hardware design of a microprocessor CPU is arranged so that a small, or very large, system can be configured around the CPU as the application demands. The internal CPU architecture, as well as the resultant machine level code that operates that architecture, is comprehensive but as flexible as possible.

The prime use of a microprocessor is to fetch data, perform extensive calculations on that data, and store those calculations in a mass storage device or display the results for human use. The programs used by the microprocessor are stored in the mass storage device and loaded into RAM as the user directs. A few microprocessor programs are stored in ROM. The ROM-based programs are primarily small fixed programs that operate peripherals and other fixed devices that are connected to the system. The design of the microprocessor is driven by the desire to make it as expandable as possible, in the expectation of commercial success in the marketplace.

Microcontrollers

Figure 1.2 shows the block diagram of a typical *microcontroller,* which is a true computer on a chip. The design incorporates all of the features found in a microprocessor CPU: ALU, PC, SP, and registers. It also has added the other features needed to make a complete computer: ROM, RAM, parallel I/O, serial I/O, counters, and a clock circuit.

Like the microprocessor, a microcontroller is a general-purpose device, but one which is meant to fetch data, perform limited calculations on that data, and control its

FIGURE 1.2 A Block Diagram of a Microcontroller

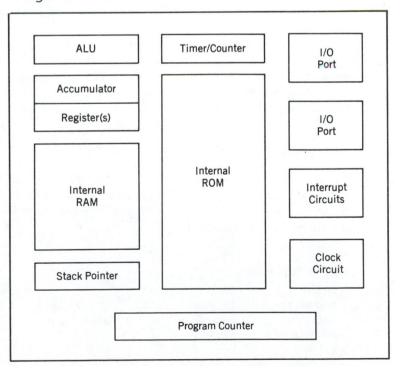

environment based on those calculations. The prime use of a microcontroller is to control the operation of a machine using a fixed program that is stored in ROM and that does not change over the lifetime of the system.

The design approach of the microcontroller mirrors that of the microprocessor: make a single design that can be used in as many applications as possible in order to sell, hopefully, as many as possible. The microprocessor design accomplishes this goal by having a very flexible and extensive repertoire of multi-byte instructions. These instructions work in a hardware configuration that enables large amounts of memory and I/O to be connected to address and data bus pins on the integrated circuit package. Much of the activity in the microprocessor has to do with moving code and data words to and from *external* memory to the CPU. The architecture features working registers that can be programmed to take part in the memory access process, and the instruction set is aimed at expediting this activity in order to improve throughput. The pins that connect the microprocessor to external memory are unique, each having a single function. Data is handled in byte, or larger, sizes.

The microcontroller design uses a much more limited set of single- and double-byte instructions that are used to move code and data from *internal* memory to the ALU. Many instructions are coupled with pins on the integrated circuit package; the pins are "programmable"—that is, capable of having several different functions depending upon the wishes of the programmer.

The microcontroller is concerned with getting data from and to its own pins; the architecture and instruction set are optimized to handle data in bit and byte size.

Comparing Microprocessors and Microcontrollers

The contrast between a microcontroller and a microprocessor is best exemplified by the fact that most microprocessors have many operational codes (opcodes) for moving data from external memory to the CPU; microcontrollers may have one, or two. Microprocessors may have one or two types of bit-handling instructions; microcontrollers will have many.

To summarize, the microprocessor is concerned with rapid movement of code and data from external addresses to the chip; the microcontroller is concerned with rapid movement of bits within the chip. The microcontroller can function as a computer with the addition of no external digital parts; the microprocessor must have many additional parts to be operational.

The Z80 and the 8051

To see the differences in concept between a microprocessor and a microcontroller, in the following table we will examine the pin configurations, architecture, and instruction sets for a very popular 8-bit microprocessor, the Zilog Z80, and an equally ubiquitous microcontroller, the 8-bit Intel 8051:

	Z80	**8051**
Pin Configurations		
Total pins	40	40
Address pins	16 (fixed)	16
Data pins	8 (fixed)	8
Interrupt pins	2 (fixed)	2
I/O pins	0	32

Continued

	Z80	**8051**
Architecture		
8-bit registers	20	34
16-bit registers	4	2
Stack size	64K	128
Internal ROM	0	4K bytes
Internal RAM	0	128 bytes
External memory	64K	128K bytes
Flags	6	4
Timers	0	2
Parallel port	0	4
Serial port	0	1
Instruction Sets (types/variations)		
External moves	4/14	2/6
Block moves	2/4	0
Bit manipulate	4/4	12/12
Jump on bit	0	3/3
Stack	3/15	2/2
Single byte	203	49
Multi-byte	490	62

Note that the point here is not to show that one design is "better" than the other; the two designs are intended to be used for different purposes and in different ways. For example, the Z80 has a very rich instruction set. The penalty that is paid for this abundance is the number of multi-byte instructions needed, some 71 percent of the total number. Each byte of a multi-byte instruction must be fetched from program memory, and each fetch takes time; this results in longer program byte counts and slower execution time versus single-byte instructions. The 8051 has a 62 percent multi-byte instruction content; the 8051 program is more compact and will run faster to accomplish similar tasks.

The disadvantage of using a "lean" instruction set as in the 8051 is increased programmer effort (expense) to write code; this disadvantage can be overcome when writing large programs by the use of high-level languages such as BASIC and C, both of which are popular with 8051 system developers. The price paid for reducing programmer time (there is *always* a price) is the size of the program generated.

A Microcontroller Survey

Markets for microcontrollers can run into millions of units per application. At these volumes the microcontroller is a commodity item and must be optimized so that cost is at a minimum. Semiconductor manufacturers have produced a mind-numbing array of designs that would seem to meet almost any need. Some of the chips listed in this section are no longer in regular production, most are current, and a few are best termed "smokeware": the dreams of an aggressive marketing department.

Four-Bit Microcontrollers

In a commodity chip, expense is represented more by the volume of the package and the number of pins it has than the amount of silicon inside. To minimize pin count and package size, it is necessary that the basic data word-bit count be held to a minimum, while still enabling useful intelligence to be implemented.

Although 4 bits, in this era of 64-bit "maximicros," may seem somewhat ludicrous, one must recall that the original 4004 was a 4-bit device, and all else followed. Indeed, in terms of production numbers, the 4-bit microcontroller is today the most popular micro made. The following table lists representative models from major manufacturers' data books. Many of these designs have been licensed to other vendors.

Manufacturer : Model	Pins : I/O	Counters	RAM (bytes)	ROM (bytes)	Other Features
Hitachi : HMCS40	28 : 10	—	32	512	10-bit ROM
National : COP420	28 : 23	1	64	1K	Serial bit I/O
OKI : MSM6411	16 : 11	—	32	1K	
TI : TMS 1000	28 : 23	—	64	1K	LED display
Toshiba : TLCS47	42 : 35	2	128	2K	Serial bit I/O

These 4-bit microcontrollers are generally intended for use in large volumes as true 1-chip computers; expanding external memory, while possible, would negate the cost advantage desired. Typical applications consist of appliances and toys; worldwide volumes run into the tens of millions.

Eight-Bit Microcontrollers

Eight-bit microcontrollers represent a transition zone between the dedicated, high volume, 4-bit microcontrollers, and the high performance, 16- and 32-bit units that will conclude this chapter.

Eight bits has proven to be a very useful word size for small computing tasks. Capable of 256 decimal values, or quarter-percent resolution, the 1-byte word is adequate for many control and monitoring applications. Serial ASCII data is also stored in byte sizes, making 8 bits the natural choice for data communications. Most integrated circuit memories and many logic functions are arranged in an 8-bit configuration that interfaces easily to data buses of 8 bits.

Application volumes for 8-bit microcontrollers may be as high as the 4-bit models, or they may be very low. Application sophistication can also range from simple appliance control to high-speed machine control and data collection. For these reasons, the microcontroller vendors have established extensive "families" of similar models. All feature a common language, but differ in the amount of internal ROM, RAM, and other cost-sensitive features. Often the memory can be expanded to include off-chip ROM and RAM; in some cases, the microcontroller has no on-board ROM at all, or the ROM is an Electrically Reprogrammable Read Only Memory (EPROM).

The purpose of this diversity is to offer the designer a menu of similar devices that can solve almost any problem. The ROMless or EPROM versions can be used by the designer to prototype the application, and then the designer can order the ROM version in large quantities from the factory. Many times the ROM version is never used. The designer makes the ROMless or EPROM design sufficiently general so that one configuration may be used many times, or because production volumes never justify the cost of a factory ROM implementation. As a further enticement for the buyer, some families have members with fewer external pins to shrink the package and the cost; others have special features such as analog-to-digital (A/D) and digital-to-analog (D/A) converters on the chip.

The 8-bit arena is crowded with capable and cleverly designed contenders; this is the growth segment of the market and the manufacturers are responding vigorously to the marketplace. The following table lists the generic family name for each chip, but keep in mind that ROMless, EPROM, and reduced pin-count members of the family are also available. Each entry in the table has many variations; the total number of configurations available exceeds a total of eighty types for the eleven model numbers listed.

Manufacturer: Model	Pins: I/O	Counters	RAM (bytes)	ROM (bytes)	Other Features
Intel: 8048	40:27	1	64	1K	External memory to 8K
Intel: 8051	40:32	2	128	4K	External memory to 128K; serial port
National: COP820	28:24	1	64	1K	Serial bit I/O
Motorola: 6805	28:20	1	64	1K	
Motorola: 68HC11	52:40	2	256	8K	Serial ports; A/D; watch dog timer (WDT)
Rockwell: 6500/1	40:32	1	64	2K	
Signetics: 87C552	68:48	3	256	8K	Serial port; A/D; WDT
TI: TMS7500	40:32	1	128	2K	External memory to 64K
TI: TMS370C050	68:55	2	256	4K	External memory to 112K; A/D; serial ports; WDT
Zilog: Z8	40:32	2	128	2K	External memory to 124K; serial port
Zilog: Z8820	44:40	2	272	8K	External memory to 128K; serial port

▷ CAUTION

Not all of the pins can be used for general-purpose I/O and addressing external memory *at the same time.* The sales literature should be read with some care to see how many of the pins have more than one function. Inspection of the table shows that the designers made tradeoffs: external memory addressing for extra on-chip functions. Generally, the ability to expand memory off of the chip implies that a ROMless family member is available for use in limited production numbers where the expense of factory programming can be avoided. Lack of this feature implies that the chip is meant for high production volumes where the expense of factory-programmed parts can be amortized over a large number of devices.

Sixteen-Bit Microcontrollers

Eight-bit microcontrollers can be used in a variety of applications that involve limited calculations and relatively simple control strategies. As the requirement for faster response and more sophisticated calculations grows, the 8-bit designs begin to hit a limit inherent with byte-wide data words. One solution is to increase clock speeds; another is to increase the size of the data word. Sixteen-bit microcontrollers have evolved to solve high-speed control problems of the type that might typically be confronted in the control of servomechanisms, such as robot arms, or for digital signal processing (DSP) applications.

The designs become much more focused on these types of real-time problems; some generality is lost, but the vendors still try to hit as many marketing targets as they can. The following table lists only three contenders. Intel has recently begun vigorously marketing the 8096 family. Other vendors are expected to appear as this market segment grows in importance.

Manufacturer : Model	Pins : I/O	Counters	RAM (bytes)	ROM (bytes)	Other Features
Hitachi : H8/532	84 : 65	5	1K	32K	External memory to 1 megabyte; serial port; A/D; pulse width modulation
Intel : 8096	68 : 40	2	232	8K	External memory to 64K; serial port; A/D; WDT; pulse width modulation
National : HPC16164	68 : 52	4	512	16K	External memory to 64K; serial port; A/D; WDT; pulse width modulation

The pulse width modulation (PWM) output is useful for controlling motor speed; it can be done using software in the 8-bit units with the usual loss of time for other tasks.

The 16- (and 32-) bit controllers have also been designed to take advantage of high-level programming languages in the expectation that very little assembly language programming will be done when employing these controllers in sophisticated applications.

Thirty-Two Bit Microcontrollers

Crossing the boundary from 16 to 32 bits involves more than merely doubling the word size of the computer. Software boundaries that separate dedicated programs from supervisory programs are also breached. Thirty two bit designs target robotics, highly intelligent instrumentation, avionics, image processing, telecommunications, automobiles, and other environments that feature application programs running under an operating system. The line between microcomputers and microcontrollers becomes very fine here.

The design emphasis now switches from on-chip features, such as RAM, ROM, timers, and serial ports, to high-speed computation features. The following table provides a general list of the capability of the Intel 80960:

HARDWARE FEATURES	SOFTWARE FEATURES
132-pin ceramic package	Efficient procedure calls
20 megahertz clock	Fault-handling capability
32-bit bus	Trace events
Floating-point unit	Global registers
512-byte instruction cache	Efficient interrupt vectors
Interrupt control	Versatile addressing

All of the functions needed for I/O, data communications, and timing and counting are done by adding other specialized chips.

This manufacturer has dubbed all of its microcontrollers "embedded controllers," a term that seems to describe the function of the 32-bit 80960 very well.

Development Systems
for Microcontrollers

What is needed to be able to apply a microcontroller to your product? That is, what package of hardware and software will allow the microcontroller to be programmed and connected to your application? A package commonly called a "development system" is required.

First, trained personnel must be available either on your technical staff or as consultants. One person who is versed in digital hardware and computer software is the minimum number.

Second, a device capable of programming EPROMs must be available to test the prototype device. Many of the microcontroller families discussed have a ROMless version, an EPROM version, or an Electrically Erasable and Programmable Read Only Memory (EEPROM) version that lets the designer debug the hardware and software prototype before committing to full-scale production. Many inexpensive EPROM programmers are sold that plug into a slot of most popular personal computers. More expensive, and more versatile, dedicated programmers are also available. An alternative to EPROMs are vendor-supplied prototype cards that allow code to be down loaded from a host computer, and the program run from RAM for debugging purposes. An EPROM will eventually have to be programmed for the production version of the microcontroller.

Finally, software is needed, along with a personal computer to host it. The minimum software package consists of a machine language assembler, which can be supplied by the microcontroller vendor or bought from independent developers. More expensive software mainly consisting of high-level language compilers and debuggers is also available.

A minimum development system, then, consists of a personal computer, a plug-in EPROM programmer, and a public-brand assembler. A more extensive system would consist of vendor-supplied dedicated computer systems with attendant high-level software packages and in-circuit emulators for hardware and software debugging. In 1990 dollars, the cost for the range of solutions outlined here is from $1000 to $10,000.

Summary

The fundamental differences between microprocessors and microcontrollers are:

□ Microprocessors are intended to be general-purpose digital computers while microcontrollers are intended to be special-purpose digital controllers.

□ Microprocessors contain a CPU, memory addressing circuits, and interrupt handling circuits. Microcontrollers have these features as well as timers, parallel and serial I/O, and internal RAM and ROM.

□ Microcontroller models vary in data size from 4 to 32 bits. Four-bit units are produced in huge volumes for very simple applications, and 8-bit units are the most versatile. Sixteen- and 32-bit units are used in high-speed control and signal processing applications.

□ Many models feature programmable pins that allow external memory to be added with the loss of I/O capability.

Questions

1. Name four major differences between a microprocessor and a microcontroller.

2. The 8051 has 40 pins on a Dual Inline Package (DIP) package, yet the comparison with the Z80 microprocessor totals 58 pins. Explain this difference.

3. Name 20 items that have a built-in microcontroller.

4. Name 10 items that should have a built-in microcontroller.

5. Name the most unusual application of a microcontroller that you have seen actually for sale.

6. Name the most likely bit size for each of the following products.

   Modem

   Printer

   Toaster

   Automobile engine control

   Robot arm

   Small ASCII data terminal

   Chess player

   House thermostat

7. Explain why ROMless versions of microcontrollers exist.

8. Name two ways to speed up digital data processing.

9. List three essential items needed to make up a development system for programming microcontrollers.

10. Search the literature and determine whether any manufacturer has announced a 64-bit microcontroller.

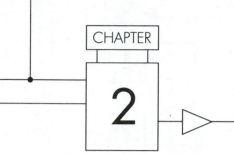

The 8051 Architecture

Chapter Outline

Introduction

The first task faced when learning to use a new computer is to become familiar with the capability of the machine. The features of the computer are best learned by studying the internal hardware design, also called the architecture of the device, to determine the type, number, and size of the registers and other circuitry.

The hardware is manipulated by an accompanying set of program instructions, or software, which is usually studied next. Once familiar with the hardware and software, the system designer can then apply the microcontroller to the problems at hand.

A natural question during this process is "What do I do with all this stuff?" Similar to attempting to write a poem in a foreign language before you have a vocabulary and rules of grammar, writing meaningful programs is not possible until you have become acquainted with both the hardware and the software of a computer.

This chapter provides a broad overview of the architecture of the 8051. In subsequent chapters, we will cover in greater detail the interaction between the hardware and the software.

8051 Microcontroller Hardware

The 8051 microcontroller actually includes a whole family of microcontrollers that have numbers ranging from 8031 to 8751 and are available in N-Channel Metal Oxide Silicon (NMOS) and Complementary Metal Oxide Silicon (CMOS) construction in a variety of

FIGURE 2.1a   8051 Block Diagram

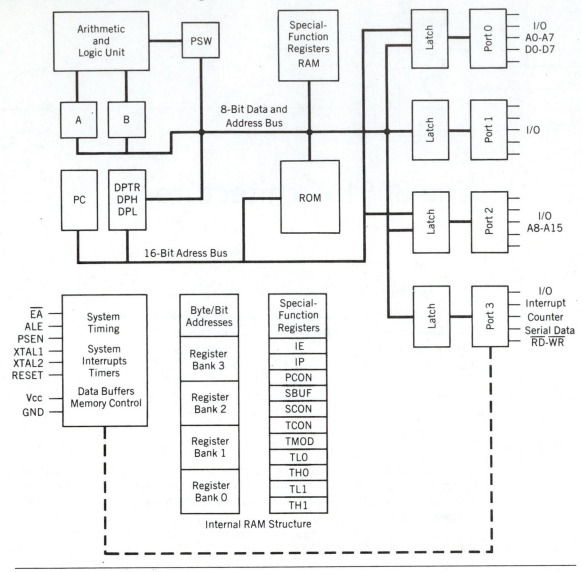

package types. An enhanced version of the 8051, the 8052, also exists with its own family of variations and even includes one member that can be programmed in BASIC. This galaxy of parts, the result of desires by the manufacturers to leave no market niche unfilled, would require many chapters to cover. In this chapter, we will study a "generic" 8051, housed in a 40-pin DIP, and direct the investigation of a particular type to the data books. The block diagram of the 8051 in Figure 2.1a shows all of the features unique to microcontrollers:

Internal ROM and RAM

I/O ports with programmable pins

Timers and counters

Serial data communication

FIGURE 2.1b   8051 Programming Model

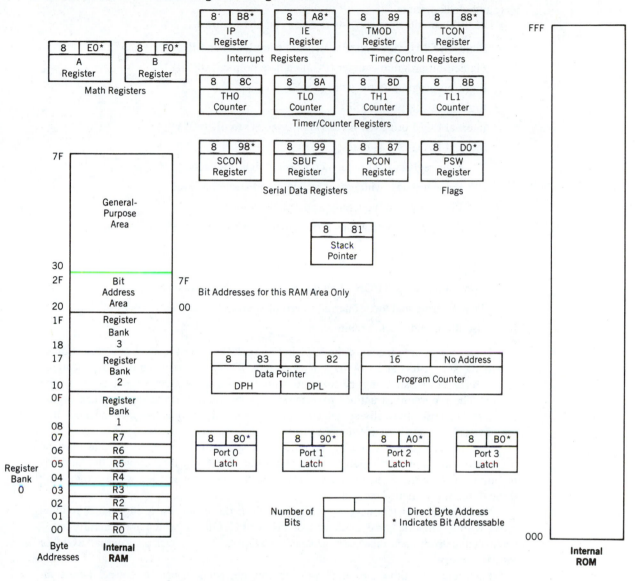

The figure also shows the usual CPU components: program counter, ALU, working registers, and clock circuits.[1]

The 8051 architecture consists of these specific features:

Eight-bit CPU with registers A (the accumulator) and B

Sixteen-bit program counter (PC) and data pointer (DPTR)

Eight-bit program status word (PSW)

Eight-bit stack pointer (SP)

Internal ROM or EPROM (8751) of 0 (8031) to 4K (8051)

Internal RAM of 128 bytes:

    Four register banks, each containing eight registers

    Sixteen bytes, which may be addressed at the bit level

    Eighty bytes of general-purpose data memory

Thirty-two input/output pins arranged as four 8-bit ports: P0–P3

Two 16-bit timer/counters: T0 and T1

Full duplex serial data receiver/transmitter: SBUF

Control registers: TCON, TMOD, SCON, PCON, IP, and IE

Two external and three internal interrupt sources

Oscillator and clock circuits

The programming model of the 8051 in Figure 2.1b shows the 8051 as a collection of 8- and 16-bit registers and 8-bit memory locations. These registers and memory locations can be made to operate using the software instructions that are incorporated as part of the design. The program instructions have to do with the control of the registers and digital data paths that are physically contained inside the 8051, as well as memory locations that are physically located outside the 8051.

The model is complicated by the number of special-purpose registers that must be present to make a microcomputer a microcontroller. A cursory inspection of the model is recommended for the first-time viewer; return to the model as needed while progressing through the remainder of the text.

Most of the registers have a specific function; those that do occupy an individual block with a symbolic name, such as A or TH0 or PC. Others, which are generally indistinguishable from each other, are grouped in a larger block, such as internal ROM or RAM memory.

Each register, with the exception of the program counter, has an internal 1-byte address assigned to it. Some registers (marked with an asterisk * in Figure 2.1b) are both byte and bit addressable. That is, the entire byte of data at such register addresses may be read or altered, or individual bits may be read or altered. Software instructions are generally able to specify a register by its address, its symbolic name, or both.

A pinout of the 8051 packaged in a 40-pin DIP is shown in Figure 2.2 with the full and abbreviated names of the signals for each pin. It is important to note that many of the

---

[1] Knowledge of the details of circuit operation that cannot be affected by any instruction or external data, while intellectually stimulating, tends to confuse the student new to the 8051. For this reason, this text will concentrate on the essential features of the 8051; the more advanced student may wish to refer to manufacturers' data books for additional information.

FIGURE 2.2 8051 DIP Pin Assignments

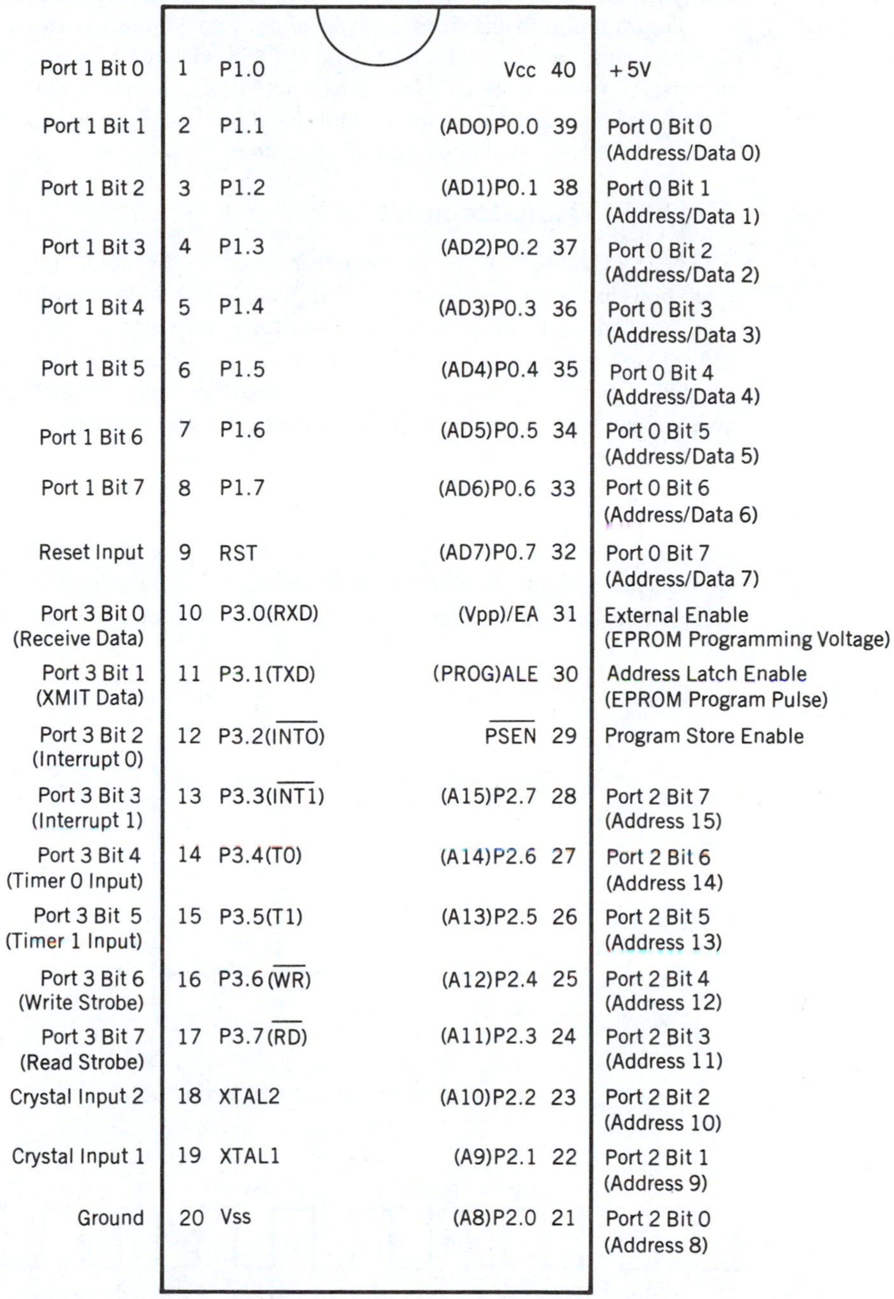

Port 1 Bit 0	1 P1.0	Vcc 40	+ 5V
Port 1 Bit 1	2 P1.1	(AD0)P0.0 39	Port 0 Bit 0 (Address/Data 0)
Port 1 Bit 2	3 P1.2	(AD1)P0.1 38	Port 0 Bit 1 (Address/Data 1)
Port 1 Bit 3	4 P1.3	(AD2)P0.2 37	Port 0 Bit 2 (Address/Data 2)
Port 1 Bit 4	5 P1.4	(AD3)P0.3 36	Port 0 Bit 3 (Address/Data 3)
Port 1 Bit 5	6 P1.5	(AD4)P0.4 35	Port 0 Bit 4 (Address/Data 4)
Port 1 Bit 6	7 P1.6	(AD5)P0.5 34	Port 0 Bit 5 (Address/Data 5)
Port 1 Bit 7	8 P1.7	(AD6)P0.6 33	Port 0 Bit 6 (Address/Data 6)
Reset Input	9 RST	(AD7)P0.7 32	Port 0 Bit 7 (Address/Data 7)
Port 3 Bit 0 (Receive Data)	10 P3.0(RXD)	(Vpp)/$\overline{EA}$ 31	External Enable (EPROM Programming Voltage)
Port 3 Bit 1 (XMIT Data)	11 P3.1(TXD)	(PROG)ALE 30	Address Latch Enable (EPROM Program Pulse)
Port 3 Bit 2 (Interrupt 0)	12 P3.2($\overline{INT0}$)	$\overline{PSEN}$ 29	Program Store Enable
Port 3 Bit 3 (Interrupt 1)	13 P3.3($\overline{INT1}$)	(A15)P2.7 28	Port 2 Bit 7 (Address 15)
Port 3 Bit 4 (Timer 0 Input)	14 P3.4(T0)	(A14)P2.6 27	Port 2 Bit 6 (Address 14)
Port 3 Bit 5 (Timer 1 Input)	15 P3.5(T1)	(A13)P2.5 26	Port 2 Bit 5 (Address 13)
Port 3 Bit 6 (Write Strobe)	16 P3.6($\overline{WR}$)	(A12)P2.4 25	Port 2 Bit 4 (Address 12)
Port 3 Bit 7 (Read Strobe)	17 P3.7($\overline{RD}$)	(A11)P2.3 24	Port 2 Bit 3 (Address 11)
Crystal Input 2	18 XTAL2	(A10)P2.2 23	Port 2 Bit 2 (Address 10)
Crystal Input 1	19 XTAL1	(A9)P2.1 22	Port 2 Bit 1 (Address 9)
Ground	20 Vss	(A8)P2.0 21	Port 2 Bit 0 (Address 8)

Note: Alternate functions are shown below the port name (in parentheses). Pin numbers and pin names are shown inside the DIP package.

pins are used for more than one function (the alternate functions are shown in parentheses in Figure 2.2). Not all of the possible 8051 features may be used *at the same time*.

Programming instructions or physical pin connections determine the use of any multi-function pins. For example, port 3 bit 0 (abbreviated P3.0) may be used as a general-purpose I/O pin, or as an input (RXD) to SBUF, the serial data receiver register. The system designer decides which of these two functions is to be used and designs the hard-ware and software affecting that pin accordingly.

The 8051 Oscillator and Clock

The heart of the 8051 is the circuitry that generates the clock pulses by which all internal operations are synchronized. Pins XTAL1 and XTAL2 are provided for connecting a reso-nant network to form an oscillator. Typically, a quartz crystal and capacitors are em-ployed, as shown in Figure 2.3. The crystal frequency is the basic internal clock fre-quency of the microcontroller. The manufacturers make available 8051 designs that can run at specified maximum and *minimum* frequencies, typically 1 megahertz to 16 mega-hertz. Minimum frequencies imply that some internal memories are dynamic and must always operate above a minimum frequency, or data will be lost.

Communication needs often dictate the frequency of the oscillator due to the require-ment that internal counters must divide the basic clock rate to yield standard communica-tion bit per second (baud) rates. If the basic clock frequency is not divisible without a remainder, then the resulting communication frequency is not standard.

FIGURE 2.3   Oscillator Circuit and Timing

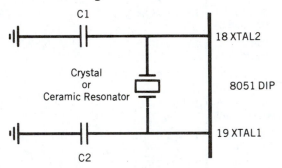

Crystal or Ceramic Resonator Oscillator Circuit

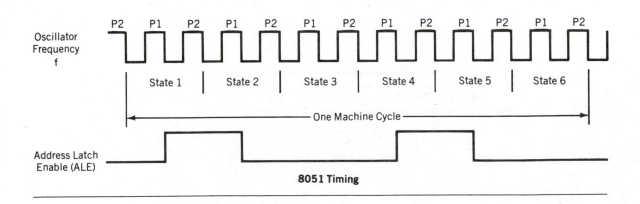

8051 Timing

Ceramic resonators may be used as a low-cost alternative to crystal resonators. However, decreases in frequency stability and accuracy make the ceramic resonator a poor choice if high-speed serial data communication with other systems, or critical timing, is to be done.

The oscillator formed by the crystal, capacitors, and an on-chip inverter generates a pulse train at the frequency of the crystal, as shown in Figure 2.3.

The clock frequency, f, establishes the smallest interval of time within the microcontroller, called the pulse, P, time. The smallest interval of time to accomplish any simple instruction, or part of a complex instruction, however, is the machine cycle. The machine cycle is itself made up of six states. A state is the basic time interval for discrete operations of the microcontroller such as fetching an opcode byte, decoding an opcode, executing an opcode, or writing a data byte. Two oscillator pulses define each state.

Program instructions may require one, two, or four machine cycles to be executed, depending on the type of instruction. Instructions are fetched and executed by the microcontroller automatically, beginning with the instruction located at ROM memory address 0000h at the time the microcontroller is first reset.

To calculate the time any particular instruction will take to be executed, find the number of cycles, C, from the list in Appendix A. The time to execute that instruction is then found by multiplying C by 12 and dividing the product by the crystal frequency:

$$\text{Tinst} = \frac{C \times 12d}{\text{crystal frequency}}$$

For example, if the crystal frequency is 16 megahertz, then the time to execute an ADD A, R1 one-cycle instruction is .75 microseconds. A 12 megahertz crystal yields the convenient time of one microsecond per cycle. An 11.0592 megahertz crystal, while seemingly an odd value, yields a cycle frequency of 921.6 kilohertz, which can be divided evenly by the standard communication baud rates of 19200, 9600, 4800, 2400, 1200, and 300 hertz.

Program Counter and Data Pointer

The 8051 contains two 16-bit registers: the program counter (PC) and the data pointer (DPTR). Each is used to hold the address of a byte in memory.

Program instruction bytes are fetched from locations in memory that are addressed by the PC. Program ROM may be on the chip at addresses 0000h to 0FFFh, external to the chip for addresses that exceed 0FFFh, or totally external for all addresses from 0000h to FFFFh. The PC is automatically incremented after every instruction byte is fetched and may also be altered by certain instructions. The PC is the only register that does not have an internal address.

The DPTR register is made up of two 8-bit registers, named DPH and DPL, that are used to furnish memory addresses for internal and external code access and external data access. The DPTR is under the control of program instructions and can be specified by its 16-bit name, DPTR, or by each individual byte name, DPH and DPL. DPTR does not have a single internal address; DPH and DPL are each assigned an address.

A and B CPU Registers

The 8051 contains 34 general-purpose, or working, registers. Two of these, registers A and B, comprise the mathematical core of the 8051 central processing unit (CPU). The other 32 are arranged as part of internal RAM in four banks, B0–B3, of eight registers each, named R0 to R7.

The A (accumulator) register is the most versatile of the two CPU registers and is used for many operations, including addition, subtraction, integer multiplication and division, and Boolean bit manipulations. The A register is also used for all data transfers between the 8051 and any external memory. The B register is used with the A register for multiplication and division operations and has no other function other than as a location where data may be stored.

Flags and the Program Status Word (PSW)

Flags are 1-bit registers provided to store the results of certain program instructions. Other instructions can test the condition of the flags and make decisions based upon the flag states. In order that the flags may be conveniently addressed, they are grouped inside the program status word (PSW) and the power control (PCON) registers.

The 8051 has four math flags that respond automatically to the outcomes of math operations and three general-purpose user flags that can be set to 1 or cleared to 0 by the programmer as desired. The math flags include carry (C), auxiliary carry (AC), overflow (OV), and parity (P). User flags are named F0, GF0, and GF1; they are general-purpose flags that may be used by the programmer to record some event in the program. Note that all of the flags can be set and cleared by the programmer at will. The math flags, however, are also affected by math operations.

The program status word is shown in Figure 2.4. The PSW contains the math flags, user program flag F0, and the register select bits that identify which of the four general-purpose register banks is currently in use by the program. The remaining two user flags, GF0 and GF1, are stored in PCON, which is shown in Figure 2.13.

FIGURE 2.4   PSW Program Status Word Register

7	6	5	4	3	2	1	0
CY	AC	F0	RS1	RS0	OV	—	P

THE PROGRAM STATUS WORD (PSW) SPECIAL FUNCTION REGISTER

Bit	Symbol	Function
7	CY	Carry flag; used in arithmetic, JUMP, ROTATE, and BOOLEAN instructions
6	AC	Auxilliary carry flag; used for BCD arithmetic
5	F0	User flag 0
4	RS1	Register bank select bit 1
3	RS0	Register bank select bit 0

RS1	RS0	
0	0	Select register bank 0
0	1	Select register bank 1
1	0	Select register bank 2
1	1	Select register bank 3

Bit	Symbol	Function
2	OV	Overflow flag; used in arithmetic instructions
1	—	Reserved for future use
0	P	Parity flag; shows parity of register A: 1 = Odd Parity

Bit addressable as PSW.0 to PSW.7

Detailed descriptions of the math flag operations will be discussed in chapters that cover the opcodes that affect the flags. The user flags can be set or cleared using data move instructions covered in Chapter 3.

Internal Memory

A functioning computer must have memory for program code bytes, commonly in ROM, and RAM memory for variable data that can be altered as the program runs. The 8051 has internal RAM and ROM memory for these functions. Additional memory can be added externally using suitable circuits.

Unlike microcontrollers with Von Neumann architectures, which can use a *single* memory address for either program code or data, *but not for both*, the 8051 has a Harvard architecture, which uses the *same address,* in *different* memories, for code and data. Internal circuitry accesses the correct memory based upon the nature of the operation in progress.

Internal RAM

The 128-byte internal RAM, which is shown generally in Figure 2.1 and in detail in Figure 2.5, is organized into three distinct areas:

1. Thirty-two bytes from address 00h to 1Fh that make up 32 working registers organized as four banks of eight registers each. The four register banks are numbered 0 to 3 and are made up of eight registers named R0 to R7. Each register can be addressed by name (when its bank is selected) or by its RAM address. Thus R0 of bank 3 is R0 (if bank 3 is currently selected) or address 18h (whether bank 3 is selected or not). Bits RS0 and RS1 in the PSW determine which bank of registers is currently in use at any time when the program is running. Register banks not selected can be used as general-purpose RAM. Bank 0 is selected upon reset.

2. A *bit*-addressable area of 16 bytes occupies RAM *byte* addresses 20h to 2Fh, forming a total of 128 addressable bits. An addressable bit may be specified by its *bit* address of 00h to 7Fh, or 8 bits may form any *byte* address from 20h to 2Fh. Thus, for example, bit address 4Fh is also bit 7 of byte address 29h. Addressable bits are useful when the program need only remember a binary event (switch on, light off, etc.). Internal RAM is in short supply as it is, so why use a byte when a bit will do?

3. A general-purpose RAM area above the bit area, from 30h to 7Fh, addressable as bytes.

The Stack and the Stack Pointer

The stack refers to an area of internal RAM that is used in conjunction with certain opcodes to store and retrieve data quickly. The 8-bit stack pointer (SP) register is used by the 8051 to hold an internal RAM address that is called the "top of the stack." The address held in the SP register is the location in internal RAM where the last byte of data was stored by a stack operation.

When data is to be placed on the stack, the SP increments *before* storing data on the stack so that the stack grows *up* as data is stored. As data is retrieved from the stack, the byte is read from the stack, and then the SP decrements to point to the next available byte of stored data.

FIGURE 2.5 Internal RAM Organization

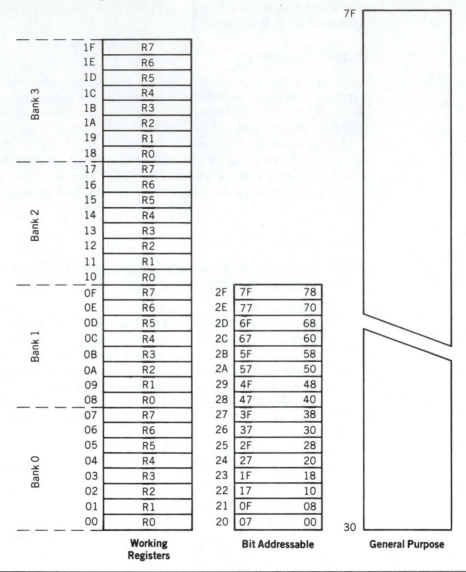

Note: Byte addresses are shown to the left; bit addresses registers are shown inside a location.

Operation of the stack and the SP is shown in Figure 2.6. The SP is set to 07h when the 8051 is reset and can be changed to any internal RAM address by the programmer.

The stack is limited in height to the size of the internal RAM. The stack has the potential (if the programmer is not careful to limit its growth) to overwrite valuable data in the register banks, bit-addressable RAM, and scratch-pad RAM areas. The programmer is responsible for making sure the stack does not grow beyond pre-defined bounds!

The stack is normally placed high in internal RAM, by an appropriate choice of the number placed in the SP register, to avoid conflict with the register, bit, and scratch-pad internal RAM areas.

FIGURE 2.6 Stack Operation

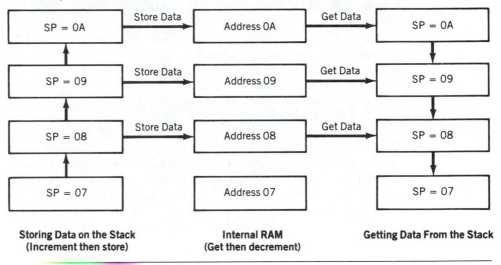

| Storing Data on the Stack (Increment then store) | Internal RAM (Get then decrement) | Getting Data From the Stack |

Special Function Registers

The 8051 operations that do not use the internal 128-byte RAM addresses from 00h to 7Fh are done by a group of specific internal registers, each called a special-function register (SFR), which may be addressed much like internal RAM, using addresses from 80h to FFh.

Some SFRs (marked with an asterisk * in Figure 2.1b) are also bit addressable, as is the case for the bit area of RAM. This feature allows the programmer to change only what needs to be altered, leaving the remaining bits in that SFR unchanged.

Not all of the addresses from 80h to FFh are used for SFRs, and attempting to use an address that is not defined, or "empty," results in unpredictable results. In Figure 2.1b, the SFR addresses are shown in the upper right corner of each block. The SFR names and equivalent internal RAM addresses are given in the following table:

NAME	FUNCTION	INTERNAL RAM ADDRESS (HEX)
A	Accumulator	0E0
B	Arithmetic	0F0
DPH	Addressing external memory	83
DPL	Addressing external memory	82
IE	Interrupt enable control	0A8
IP	Interrupt priority	0B8
P0	Input/output port latch	80
P1	Input/output port latch	90
P2	Input/output port latch	A0
P3	Input/output port latch	0B0
PCON	Power control	87
PSW	Program status word	0D0
SCON	Serial port control	98
SBUF	Serial port data buffer	99

Continued

NAME	FUNCTION	INTERNAL RAM ADDRESS (HEX)
		Continued
SP	Stack pointer	81
TMOD	Timer/counter mode control	89
TCON	Timer/counter control	88
TL0	Timer 0 low byte	8A
TH0	Timer 0 high byte	8C
TL1	Timer 1 low byte	8B
TH1	Timer 1 high byte	8D

Note that the PC is not part of the SFR and has no internal RAM address.

SFRs are named in certain opcodes by their functional names, such as A or TH0, and are referenced by other opcodes by their addresses, such as 0E0h or 8Ch. Note that *any* address used in the program *must* start with a number; thus address E0h for the A SFR begins with 0. Failure to use this number convention will result in an assembler error when the program is assembled.

Internal ROM

The 8051 is organized so that data memory and program code memory can be in two entirely different physical memory entities. *Each* has the same address ranges.

The structure of the internal RAM has been discussed previously. A corresponding block of internal program code, contained in an internal ROM, occupies code address space 0000h to 0FFFh. The PC is ordinarily used to address program code bytes from addresses 0000h to FFFFh. Program addresses higher than 0FFFh, which exceed the internal ROM capacity, will cause the 8051 to automatically fetch code bytes from external program memory. Code bytes can also be fetched exclusively from an external memory, addresses 0000h to FFFFh, by connecting the external access pin ($\overline{EA}$ pin 31 on the DIP) to ground. The PC does not care where the code is; the circuit designer decides whether the code is found totally in internal ROM, totally in external ROM, or in a combination of internal and external ROM.

Input/Output Pins, Ports, and Circuits

One major feature of a microcontroller is the versatility built into the input/output (I/O) circuits that connect the 8051 to the outside world. As noted in Chapter 1, microprocessor designs must add additional chips to interface with external circuitry; this ability is built into the microcontroller.

To be commercially viable, the 8051 had to incorporate as many functions as were technically and economically feasible. The main constraint that limits numerous functions is the number of pins available to the 8051 circuit designers. The DIP has 40 pins, and the success of the design in the marketplace was determined by the flexibility built into the use of these pins.

For this reason, 24 of the pins may each be used for one of two entirely different functions, yielding a total pin configuration of 64. The function a pin performs at any given instant depends, first, upon what is physically connected to it and, then, upon what software commands are used to "program" the pin. Both of these factors are under the complete control of the 8051 programmer and circuit designer.

Given this pin flexibility, the 8051 may be applied simply as a single component with I/O only, or it may be expanded to include additional memory, parallel ports, and serial data communication by using the alternate pin assignments. The key to programming an alternate pin function is the port pin circuitry shown in Figure 2.7.

Each port has a D-type output latch for each pin. The SFR for each port is made up of these eight latches, which can be addressed at the SFR address for that port. For instance, the eight latches for port 0 are addressed at location 80h; port 0 pin 3 is bit 2 of the P0 SFR. The port latches should not be confused with the port pins; the data on the latches does *not* have to be the same as that on the pins.

The two data paths are shown in Figure 2.7 by the circuits that read the latch or pin data using two entirely separate buffers. The top buffer is enabled when latch data is read, and the lower buffer, when the pin state is read. The status of each latch may be read from a latch buffer, while an input buffer is connected directly to each pin so that the pin status may be read independently of the latch state.

Different opcodes access the latch or pin states as appropriate. Port operations are determined by the manner in which the 8051 is connected to external circuitry.

Programmable port pins have completely different alternate functions. The configuration of the control circuitry between the output latch and the port pin determines the nature of any particular port pin function. An inspection of Figure 2.7 reveals that only port 1 cannot have alternate functions; ports 0, 2, and 3 can be programmed.

The ports are not capable of driving loads that require currents in the tens of milliamperes (mA). As previously mentioned, the 8051 has many family members, and many are fabricated in varying technologies. An example range of logic-level currents, voltages, and total device power requirements is given in the following table:

Parameter	V_{oh}	I_{oh}	V_{ol}	I_{ol}	V_{il}	I_{il}	V_{ih}	I_{ih}	P_t
CMOS	2.4 V	$-60\ \mu A$	.45 V	1.6 mA	.9 V	$|10\ \mu A|$	1.9 V	$|10\ \mu A|$	50 mW
NMOS	2.4 V	$-80\ \mu A$	.45 V	1.6 mA	.8 V	$-800\ \mu A$	2.0 V	$10\ \mu A$	800 mW

These figures tell us that driving more than two LSTTL inputs degrades the noise immunity of the ports and that careful attention must be paid to buffering the ports when they must drive currents in excess of those listed. Again, one must refer to the manufacturers' data books when designing a "real" application.

Port 0

Port 0 pins may serve as inputs, outputs, or, when used together, as a bi-directional low-order address and data bus for external memory. For example, when a pin is to be used as an input, a 1 *must be* written to the corresponding port 0 latch by the program, thus turning both of the output transistors off, which in turn causes the pin to "float" in a high-impedance state, and the pin is essentially connected to the input buffer.

When used as an output, the pin latches that are programmed to a 0 will turn on the lower FET, grounding the pin. All latches that are programmed to a 1 still float; thus, external pullup resistors will be needed to supply a logic high when using port 0 as an output.

When port 0 is used as an address bus to external memory, internal control signals switch the address lines to the gates of the Field Effect Transistories (FETs). A logic 1 on an address bit will turn the upper FET on and the lower FET off to provide a logic high at the pin. When the address bit is a zero, the lower FET is on and the upper FET off to

FIGURE 2.7 Port Pin Circuits

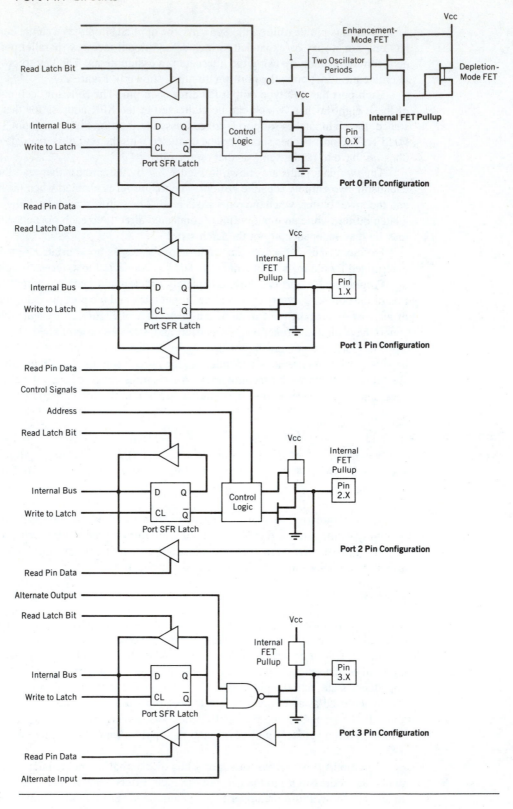

provide a logic low at the pin. After the address has been formed and latched into external circuits by the Address Latch Enable (ALE) pulse, the bus is turned around to become a data bus. Port 0 now reads data from the external memory and must be configured as an input, so a logic 1 is automatically written by internal control logic to all port 0 latches.

Port 1

Port 1 pins have no dual functions. Therefore, the output latch is connected directly to the gate of the lower FET, which has an FET circuit labeled "Internal FET Pullup" as an active pullup load.

Used as an input, a 1 is written to the latch, turning the lower FET off; the pin and the input to the pin buffer are pulled high by the FET load. An external circuit can overcome the high impedance pullup and drive the pin low to input a 0 or leave the input high for a 1.

If used as an output, the latches containing a 1 can drive the input of an external circuit high through the pullup. If a 0 is written to the latch, the lower FET is on, the pullup is off, and the pin can drive the input of the external circuit low.

To aid in speeding up switching times when the pin is used as an output, the internal FET pullup has another FET in parallel with it. The second FET is turned on for two oscillator time periods during a low-to-high transition on the pin, as shown in Figure 2.7. This arrangement provides a low impedance path to the positive voltage supply to help reduce rise times in charging any parasitic capacitances in the external circuitry.

Port 2

Port 2 may be used as an input/output port similar in operation to port 1. The alternate use of port 2 is to supply a high-order address byte in conjunction with the port 0 low-order byte to address external memory.

Port 2 pins are momentarily changed by the address control signals when supplying the high byte of a 16-bit address. Port 2 latches remain stable when external memory is addressed, as they do not have to be turned around (set to 1) for data input as is the case for port 0.

Port 3

Port 3 is an input/output port similar to port 1. The input and output functions can be programmed under the control of the P3 latches or under the control of various other special function registers. The port 3 alternate uses are shown in the following table:

PIN	ALTERNATE USE	SFR
P3.0–RXD	Serial data input	SBUF
P3.1–TXD	Serial data output	SBUF
P3.2–$\overline{INT0}$	External interrupt 0	TCON.1
P3.3–$\overline{INT1}$	External interrupt 1	TCON.3
P3.4–T0	External timer 0 input	TMOD
P3.5–T1	External timer 1 input	TMOD
P3.6–$\overline{WR}$	External memory write pulse	—
P3.7–$\overline{RD}$	External memory read pulse	—

Unlike ports 0 and 2, which can have external addressing functions and change all eight port bits when in alternate use, each pin of port 3 may be individually programmed to be used either as I/O or as one of the alternate functions.

External Memory

The system designer is not limited by the amount of internal RAM and ROM available on chip. Two separate external memory spaces are made available by the 16-bit PC and DPTR and by different control pins for enabling external ROM and RAM chips. Internal control circuitry accesses the correct physical memory, depending upon the machine cycle state and the opcode being executed.

There are several reasons for adding external memory, particularly program memory, when applying the 8051 in a system. When the project is in the prototype stage, the expense—in time and money—of having a masked internal ROM made for each program "try" is prohibitive. To alleviate this problem, the manufacturers make available an EPROM version, the 8751, which has 4K of on-chip EPROM that may be programmed and erased as needed as the program is developed. The resulting circuit board layout will be identical to one that uses a factory-programmed 8051. The only drawbacks to the 8751 are the specialized EPROM programmers that must be used to program the non-standard 40-pin part, and the limit of "only" 4096 bytes of program code.

The 8751 solution works well if the program will fit into 4K bytes. Unfortunately, many times, particularly if the program is written in a high-level language, the program size exceeds 4K bytes, and an external program memory is needed. Again, the manufacturers provide a version for the job, the ROMless 8031. The $\overline{EA}$ pin is grounded when using the 8031, and all program code is contained in an external EPROM that may be as large as 64K bytes and that can be programmed using standard EPROM programmers.

External RAM, which is accessed by the DPTR, may also be needed when 128 bytes of internal data storage is not sufficient. External RAM, up to 64K bytes, may also be added to any chip in the 8051 family.

Connecting External Memory

Figure 2.8 shows the connections between an 8031 and an external memory configuration consisting of 16K bytes of EPROM and 8K bytes of static RAM. The 8051 accesses external RAM whenever certain program instructions are executed. External ROM is accessed whenever the $\overline{EA}$ (external access) pin is connected to ground or when the PC contains an address higher than the last address in the internal 4K bytes ROM (0FFFh). 8051 designs can thus use internal and external ROM automatically; the 8031, having no internal ROM, must have $\overline{EA}$ grounded.

Figure 2.9 shows the timing associated with an external memory access cycle. During any memory access cycle, port 0 is time multiplexed. That is, it first provides the lower byte of the 16-bit memory address, then acts as a bidirectional data bus to write or read a byte of memory data. Port 2 provides the high byte of the memory address during the entire memory read/write cycle.

The lower address byte from port 0 must be latched into an external register to save the byte. Address byte save is accomplished by the ALE clock pulse that provides the correct timing for the '373 type data latch. The port 0 pins then become free to serve as a data bus.

If the memory access is for a byte of program code in the ROM, the $\overline{PSEN}$ (program store enable) pin will go low to enable the ROM to place a byte of program code on the data bus. If the access is for a RAM byte, the $\overline{WR}$ (write) or $\overline{RD}$ (read) pins will go low, enabling data to flow between the RAM and the data bus.

The ROM may be expanded to 64K by using a 27512 type EPROM and connecting the remaining port 2 upper address lines A14–A15 to the chip.

At this time the largest static RAMs available are 32K in size; RAM can be expanded to 64K by using two 32K RAMs that are connected through address A14 of port 2. The

FIGURE 2.8 External Memory Connections

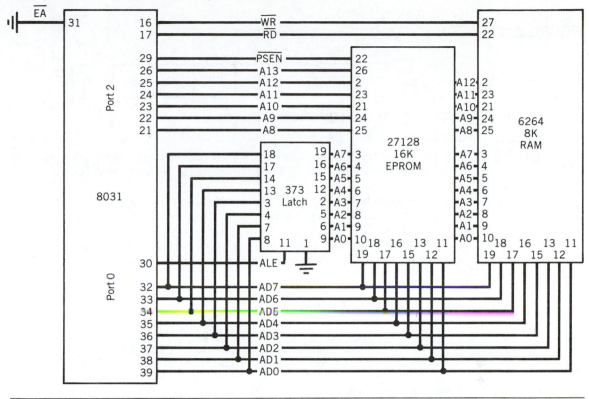

FIGURE 2.9 External Memory Timing

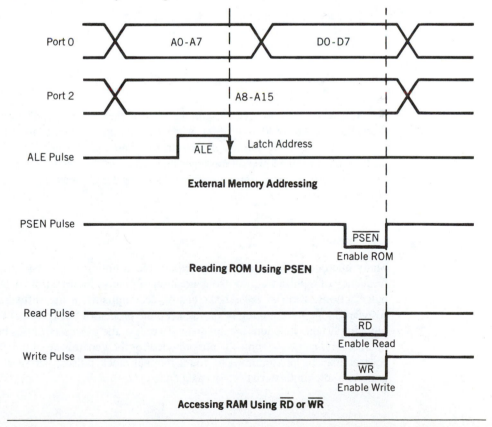

FIGURE 2.10   TCON and TMOD Function Registers

7	6	5	4	3	2	1	0
TF1	TR1	TF0	TR0	IE1	IT1	IE0	IT0

THE TIMER CONTROL (TCON) SPECIAL FUNCTION REGISTER

Bit	Symbol	Function
7	TF1	Timer 1 Overflow flag. Set when timer rolls from all ones to zero. Cleared when processor vectors to execute interrupt service routine located at program address 001Bh.
6	TR1	Timer 1 run control bit. Set to 1 by program to enable timer to count; cleared to 0 by program to halt timer. Does not reset timer.
5	TF0	Timer 0 Overflow flag. Set when timer rolls from all ones to zero. Cleared when processor vectors to execute interrupt service routine located at program address 000Bh.
4	TR0	Timer 0 run control bit. Set to 1 by program to enable timer to count; cleared to 0 by program to halt timer. Does not reset timer.
3	IE1	External interrupt 1 edge flag. Set to 1 when a high to low edge signal is received on port 3 pin 3.3 ($\overline{\text{INT1}}$). Cleared when processor vectors to interrupt service routine located at program address 0013h. Not related to timer operations.
2	IT1	External interrupt 1 signal type control bit. Set to 1 by program to enable external interrupt 1 to be triggered by a falling edge signal. Set to 0 by program to enable a low level signal on external interrupt 1 to generate an interrupt.
1	IE0	External interrupt 0 edge flag. Set to 1 when a high to low edge signal is received on port 3 pin 3.2 ($\overline{\text{INT0}}$). Cleared when processor vectors to interrupt service routine located at program address 0003h. Not related to timer operations.

Continued

first 32K RAM (0000h – 7FFFh) can then be enabled when A15 of port 2 is low, and the second 32K RAM (8000h – FFFFh) when A15 is high, by using an inverter.

Note that the $\overline{\text{WR}}$ and $\overline{\text{RD}}$ signals are alternate uses for port 3 pins 16 and 17. Also, port 0 is used for the lower address byte and data; port 2 is used for upper address bits. The use of external memory consumes many of the port pins, leaving only port 1 and parts of port 3 for general I/O.

Counters and Timers

Many microcontroller applications require the counting of external events, such as the frequency of a pulse train, or the generation of precise internal time delays between computer actions. Both of these tasks can be accomplished using software techniques, but software loops for counting or timing keep the processor occupied so that other, perhaps more important, functions are not done. To relieve the processor of this burden, two 16-bit *up* counters, named T0 and T1, are provided for the general use of the programmer. Each counter may be programmed to count internal clock pulses, acting as a timer, or programmed to count external pulses as a counter.

Bit	Symbol	Function
0	IT0	External interrupt 0 signal type control bit. Set to 1 by program to enable external interrupt 0 to be triggered by a falling edge signal. Set to 0 by program to enable a low level signal on external interrupt 0 to generate an interrupt.

Bit addressable as TCON.0 to TCON.7

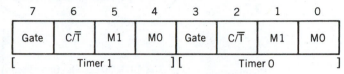

7	6	5	4	3	2	1	0
Gate	C/$\overline{T}$	M1	M0	Gate	C/$\overline{T}$	M1	M0

[Timer 1   ] [Timer 0   ]

THE TIMER MODE CONTROL (TMOD) SPECIAL FUNCTION REGISTER

Bit	Symbol	Function
7/3	Gate	OR gate enable bit which controls RUN/STOP of timer 1/0. Set to 1 by program to enable timer to run if bit TR1/0 in TCON is set and signal on external interrupt $\overline{INT1}$/0 pin is high. Cleared to 0 by program to enable timer to run if bit TR1/0 in TCON is set.
6/2	C/$\overline{T}$	Set to 1 by program to make timer 1/0 act as a counter by counting pulses from external input pins 3.5 (T1) or 3.4 (T0). Cleared to 0 by program to make timer act as a timer by counting internal frequency.
5/1	M1	Timer/counter operating mode select bit 1. Set/cleared by program to select mode.
4/0	M0	Timer/counter operating mode select bit 0. Set/cleared by program to select mode.

M1	M0	Mode
0	0	0
0	1	1
1	0	2
1	1	3

TMOD is not bit addressable

The counters are divided into two 8-bit registers called the timer low (TL0, TL1) and high (TH0, TH1) bytes. All counter action is controlled by bit states in the timer mode control register (TMOD), the timer/counter control register (TCON), and certain program instructions.

TMOD is dedicated solely to the two timers and can be considered to be two duplicate 4-bit registers, each of which controls the action of one of the timers. TCON has control bits and flags for the timers in the upper nibble, and control bits and flags for the external interrupts in the lower nibble. Figure 2.10 shows the bit assignments for TMOD and TCON.

Timer Counter Interrupts

The counters have been included on the chip to relieve the processor of timing and counting chores. When the program wishes to count a certain number of internal pulses or external events, a number is placed in one of the counters. The number represents the maximum count *less* the desired count, plus one. The counter increments from the initial number to the maximum and then rolls over to zero on the final pulse and also sets a timer flag. The flag condition may be tested by an instruction to tell the program that the count has been accomplished, or the flag may be used to interrupt the program.

FIGURE 2.11 Timer/Counter Control Logic

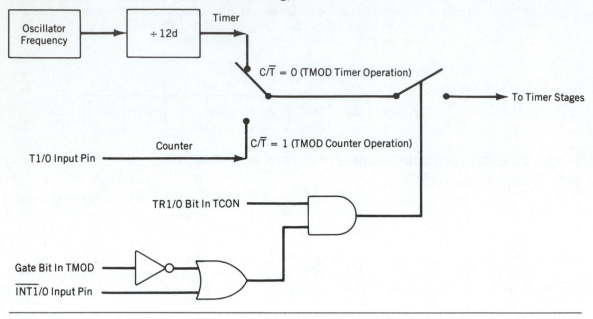

Timing

If a counter is programmed to be a timer, it will count the internal clock frequency of the 8051 oscillator divided by 12d. As an example, if the crystal frequency is 6.0 megahertz, then the timer clock will have a frequency of 500 kilohertz.

The resultant timer clock is gated to the timer by means of the circuit shown in Figure 2.11. In order for oscillator clock pulses to reach the timer, the $C/\overline{T}$ bit in the TMOD register must be set to 0 (timer operation). Bit TRX in the TCON register must be set to 1 (timer run), and the *gate* bit in the TMOD register must be 0, or external pin $\overline{INTX}$ must be a 1. In other words, the counter is configured as a timer, then the timer pulses are gated to the counter by the run bit *and* the gate bit *or* the external input bits $\overline{INTX}$.

Timer Modes of Operation

The timers may operate in any one of four modes that are determined by the mode bits, M1 and M0, in the TMOD register. Figure 2.12 shows the four timer modes.

Timer Mode 0

Setting timer X mode bits to 00b in the TMOD register results in using the THX register as an 8-bit counter and TLX as a 5-bit counter; the pulse input is divided by 32d in TL so that TH counts the original oscillator frequency reduced by a total 384d. As an example, the 6 megahertz oscillator frequency would result in a final frequency to TH of 15625 hertz. The timer flag is set whenever THX goes from FFh to 00h, or in .0164 seconds for a 6 megahertz crystal if THX starts at 00h.

Timer Mode 1

Mode 1 is similar to mode 0 except TLX is configured as a full 8-bit counter when the mode bits are set to 01b in TMOD. The timer flag would be set in .1311 seconds using a 6 megahertz crystal.

FIGURE 2.12   Timer 1 and Timer 0 Operation Modes

Pulse Input (Figure 2.11) → | TLX 5 Bits | | THX 8 Bits | | TFX | → Interrupt

Timer Mode 0 13 - Bit Timer/Counter

Pulse Input (Figure 2.11) → | TLX 8 Bits | | THX 8 Bits | | TFX | → Interrupt

Timer Mode 1 16 - Bit Timer/Counter

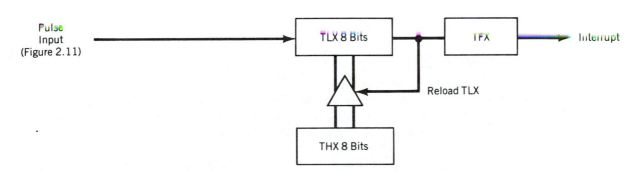

Pulse Input (Figure 2.11) → | TLX 8 Bits | | TFX | → Interrupt

Reload TLX

| THX 8 Bits |

Timer Mode 2 Auto - Reload of TL from TH

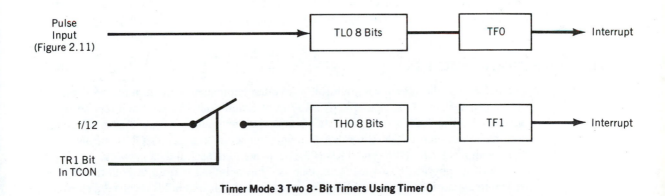

Pulse Input (Figure 2.11) → | TL0 8 Bits | | TF0 | → Interrupt

f/12 — | TH0 8 Bits | | TF1 | → Interrupt

TR1 Bit In TCON

Timer Mode 3 Two 8 - Bit Timers Using Timer 0

Timer Mode 2

Setting the mode bits to 10b in TMOD configures the timer to use only the TLX counter as an 8-bit counter. THX is used to hold a value that is loaded into TLX every time TLX overflows from FFh to 00h. The timer flag is also set when TLX overflows.

This mode exhibits an auto-reload feature: TLX will count up from the number in THX, overflow, and be initialized again with the contents of THX. For example, placing

9Ch in THX will result in a delay of exactly .0002 seconds before the overflow flag is set if a 6 megahertz crystal is used.

Timer Mode 3

Timers 0 and 1 may be programmed to be in mode 0, 1, or 2 independently of a similar mode for the other timer. This is not true for mode 3; the timers do not operate independently if mode 3 is chosen for timer 0. Placing timer 1 in mode 3 causes it to stop counting; the control bit TR1 and the timer 1 flag TF1 are then used by timer 0.

Timer 0 in mode 3 becomes two completely separate 8-bit counters. TL0 is controlled by the gate arrangement of Figure 2.11 and sets timer flag TF0 whenever it overflows from FFh to 00h. TH0 receives the timer clock (the oscillator divided by 12) under the control of TR1 only and sets the TF1 flag when it overflows.

Timer 1 may still be used in modes 0, 1, and 2, while timer 0 is in mode 3 with one important exception: *No interrupts* will be generated by timer 1 while timer 0 is using the TF1 overflow flag. Switching timer 1 to mode 3 will stop it (and hold whatever count is in timer 1). Timer 1 can be used for baud rate generation for the serial port, or any other mode 0, 1, or 2 function that does not depend upon an interrupt (or any other use of the TF1 flag) for proper operation.

Counting

The only difference between counting and timing is the source of the clock pulses to the counters. When used as a timer, the clock pulses are sourced from the oscillator through the divide-by-12d circuit. When used as a counter, pin T0 (P3.4) supplies pulses to counter 0, and pin T1 (P3.5) to counter 1. The C/$\overline{\text{T}}$ bit in TMOD must be set to 1 to enable pulses from the TX pin to reach the control circuit shown in Figure 2.11.

The input pulse on TX is sampled during P2 of state 5 every machine cycle. A change on the input from high to low between samples will increment the counter. Each high and low state of the input pulse must thus be held constant for at least one machine cycle to ensure reliable counting. Since this takes 24 pulses, the maximum input frequency that can be accurately counted is the oscillator frequency divided by 24. For our 6 megahertz crystal, the calculation yields a maximum external frequency of 250 kilohertz.

Serial Data Input/Output

Computers must be able to communicate with other computers in modern multiprocessor distributed systems. One cost-effective way to communicate is to send and receive data bits serially. The 8051 has a serial data communication circuit that uses register SBUF to hold data. Register SCON controls data communication, register PCON controls data rates, and pins RXD (P3.0) and TXD (P3.1) connect to the serial data network.

SBUF is physically two registers. One is write only and is used to hold data to be transmitted *out* of the 8051 via TXD. The other is read only and holds received data *from* external sources via RXD. Both mutually exclusive registers use address 99h.

There are four programmable modes for serial data communication that are chosen by setting the SMX bits in SCON. Baud rates are determined by the mode chosen. Figure 2.13 shows the bit assignments for SCON and PCON.

Serial Data Interrupts

Serial data communication is a relatively slow process, occupying many milliseconds per data byte to accomplish. In order not to tie up valuable processor time, serial data flags are

FIGURE 2.13 SCON and PCON Function Registers

7	6	5	4	3	2	1	0
SM0	SM1	SM2	REN	TB8	RB8	TI	RI

THE SERIAL PORT CONTROL (SCON) SPECIAL FUNCTION REGISTER

Bit	Symbol	Function
7	SM0	Serial port mode bit 0. Set/cleared by program to select mode.
6	SM1	Serial port mode bit 1. Set/cleared by program to select mode.

SM0	SM1	Mode	Description
0	0	0	Shift register; baud = f/12
0	1	1	8-bit UART; baud = variable
1	0	2	9-bit UART; baud = f/32 or f/64
1	1	3	9-bit UART; baud = variable

Bit	Symbol	Function
5	SM2	Multiprocessor communications bit. Set/cleared by program to enable multiprocessor communications in modes 2 and 3. When set to 1 an interrupt is generated if bit 9 of the received data is a 1; no interrupt is generated if bit 9 is a 0. If set to 1 for mode 1, no interrupt will be generated unless a valid stop bit is received. Clear to 0 if mode 0 is in use.
4	REN	Receive enable bit. Set to 1 to enable reception; cleared to 0 to dissable reception.
3	TB8	Transmitted bit 8. Set/cleared by program in modes 2 and 3.
2	RB8	Received bit 8. Bit 8 of received data in modes 2 and 3; stop bit in mode 1. Not used in mode 0.
1	TI	Transmit interrupt flag. Set to one at the end of bit 7 time in mode 0, and at the beginning of the stop bit for other modes. Must be cleared by the program.
0	RI	Receive interrupt flag. Set to one at the end of bit 7 time in mode 0, and halfway through the stop bit for other modes. Must be cleared by the program.

Bit addressable as SCON.0 to SCON.7

7	6	5	4	3	2	1	0
SMOD	—	—	—	GF1	GF0	PD	IDL

THE POWER MODE CONTROL (PCON) SPECIAL FUNCTION REGISTER

Bit	Symbol	Function
7	SMOD	Serial baud rate modify bit. Set to 1 by program to double baud rate using timer 1 for modes 1, 2, and 3. Cleared to 0 by program to use timer 1 baud rate.
6-4	—	Not implemented.
3	GF1	General purpose user flag bit 1. Set/cleared by program.
2	GF0	General purpose user flag bit 0. Set/cleared by program.
1	PD	Power down bit. Set to 1 by program to enter power down configuration for CHMOS processors.
0	IDL	Idle mode bit. Set to 1 by program to enter idle mode configuration for CHMOS processors. PCON is not bit addressable.

included in SCON to aid in efficient data transmission and reception. Notice that data transmission is under the complete control of the program, but reception of data is unpredictable and at random times that are beyond the control of the program.

The serial data flags in SCON, TI and RI, are set whenever a data byte is transmitted (TI) or received (RI). These flags are ORed together to produce an interrupt to the program. The program must read these flags to determine which caused the interrupt and then clear the flag. This is unlike the timer flags that are cleared automatically; it is the responsibility of the programmer to write routines that handle the serial data flags.

Data Transmission

Transmission of serial data bits begins *anytime* data is written to SBUF. TI is set to a 1 when the data has been transmitted and signifies that SBUF is empty (for transmission purposes) and that another data byte can be sent. If the program fails to wait for the TI flag and overwrites SBUF while a previous data byte is in the process of being transmitted, the results will be unpredictable (a polite term for "garbage out").

Data Reception

Reception of serial data will begin *if* the receive enable bit (REN) in SCON is set to 1 for all modes. In addition, for mode 0 *only,* RI must be cleared to 0 also. Receiver interrupt flag RI is set after data has been received in all modes. Setting REN is the only direct program control that limits the reception of unexpected data; the requirement that RI also be 0 for mode 0 prevents the reception of new data until the program has dealt with the old data and reset RI.

Reception can *begin* in modes 1, 2, and 3 if RI is set when the serial stream of bits begins. RI must have been reset by the program before the *last* bit is received or the *incoming* data will be lost. Incoming data is not transferred to SBUF until the last data bit has been received so that the previous transmission can be read from SBUF while new data is being received.

Serial Data Transmission Modes

The 8051 designers have included four modes of serial data transmission that enable data communication to be done in a variety of ways and a multitude of baud rates. Modes are selected by the programmer by setting the mode bits SM0 and SM1 in SCON. Baud rates are fixed for mode 0 and variable, using timer 1 and the serial baud rate modify bit (SMOD) in PCON, for modes 1, 2, and 3.

Serial Data Mode 0—Shift Register Mode

Setting bits SM0 and SM1 in SCON to 00b configures SBUF to receive or transmit eight data bits using pin RXD for *both* functions. Pin TXD is connected to the internal shift frequency pulse source to supply shift pulses to external circuits. The shift frequency, or baud rate, is fixed at 1/12 of the oscillator frequency, the same rate used by the timers when in the timer configuration. The TXD shift clock is a square wave that is low for machine cycle states S3–S4–S5 and high for S6–S1–S2. Figure 2.14 shows the timing for mode 0 shift register data transmission.

When transmitting, data is shifted *out* of RXD; the data changes on the *falling* edge of S6P2, or one clock pulse after the *rising* edge of the output TXD shift clock. The system designer must design the external circuitry that receives this transmitted data to receive the data reliably based on this timing.

FIGURE 2.14 Shift Register Mode 0 Timing

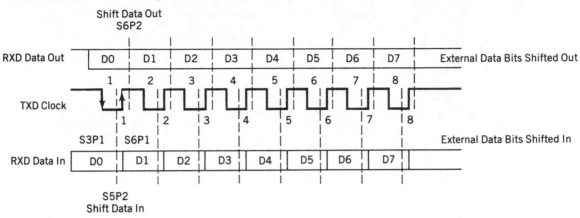

Received data comes *in* on pin RXD and should be synchronized with the shift clock produced at TXD. Data is sampled on the *falling* edge of S5P2 and shifted in to SBUF on the *rising* edge of the shift clock.

Mode 0 is intended *not* for data communication between computers, but as a high-speed serial data-collection method using discrete logic to achieve high data rates. The baud rate used in mode 0 will be much higher than standard for any reasonable oscillator frequency; for a 6 megahertz crystal, the shift rate will be 500 kilohertz.

Serial Data Mode 1—Standard UART

When SM0 and SM1 are set to 01b, SBUF becomes a 10-bit full-duplex receiver/transmitter that may receive and transmit data at the same time. Pin RXD receives all data, and pin TXD transmits all data. Figure 2.15 shows the format of a data word.

Transmitted data is sent as a start bit, eight data bits (Least Significant Bit, LSB, first), and a stop bit. Interrupt flag TI is set once all ten bits have been sent. Each bit interval is the inverse of the baud rate frequency, and each bit is maintained high or low over that interval.

Received data is obtained in the same order; reception is triggered by the falling edge of the start bit and continues if the stop bit is true (0 level) halfway through the start bit interval. This is an anti-noise measure; if the reception circuit is triggered by noise on the transmission line, the check for a low after half a bit interval should limit false data reception.

Data bits are shifted into the receiver at the programmed baud rate, and the data word will be loaded to SBUF *if* the following conditions are true: RI *must* be 0, *and* mode bit SM2 is 0 *or* the stop bit is 1 (the normal state of stop bits). RI set to 0 implies that the program has read the previous data byte and is ready to receive the next; a normal stop bit will then complete the transfer of data to SBUF regardless of the state of SM2. SM2 set to 0 enables the reception of a byte with any stop bit state, a condition which is of limited use in this mode, but very useful in modes 2 and 3. SM2 set to 1 forces reception of only "good" stop bits, an anti-noise safeguard.

Of the original ten bits, the start bit is discarded, the eight data bits go to SBUF, and the stop bit is saved in bit RB8 of SCON. RI is set to 1, indicating a new data byte has been received.

FIGURE 2.15 Standard UART Data Word

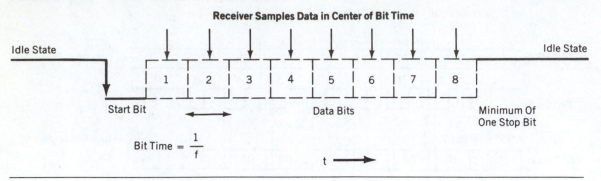

If RI is found to be set at the end of the reception, indicating that the previously received data byte has not been read by the program, or if the other conditions listed are not true, the *new* data will not be loaded and will be *lost*.

Mode 1 Baud Rates

Timer 1 is used to generate the baud rate for mode 1 by using the overflow flag of the timer to determine the baud frequency. Typically, timer 1 is used in timer mode 2 as an autoload 8-bit timer that generates the baud frequency:

$$f_{baud} = \frac{2^{SMOD}}{32d} \times \frac{\text{oscillator frequency}}{12d \times [256d - (TH1)]}$$

SMOD is the control bit in PCON and can be 0 or 1, which raises the 2 in the equation to a value of 1 or 2.

If timer 1 is not run in timer mode 2, then the baud rate is

$$f_{baud} = \frac{2^{SMOD}}{32d} \times (\text{timer 1 overflow frequency})$$

and timer 1 can be run using the internal clock or as a counter that receives clock pulses from any external source via pin T1.

The oscillator frequency is chosen to help generate both standard and nonstandard baud rates. If standard baud rates are desired, then an 11.0592 megahertz crystal could be selected. To get a standard rate of 9600 hertz then, the setting of TH1 may be found as follows:

$$TH1 = 256d - \left(\frac{2^0}{32d} \times \frac{11.0592 \times 10^6}{12 \times 9600d} \right) = 253.0000d = 0FDh$$

if SMOD is cleared to 0.

Serial Data Mode 2—Multiprocessor Mode

Mode 2 is similar to mode 1 except 11 bits are transmitted: a start bit, nine data bits, and a stop bit, as shown in Figure 2.16. The ninth data bit is gotten from bit TB8 in SCON during transmit and stored in bit RB8 of SCON when data is received. Both the start and stop bits are discarded.

The baud rate is programmed as follows:

$$f_{baud2} = \frac{2^{SMOD}}{64d} \times \text{oscillator frequency}$$

FIGURE 2.16 Multiprocessor Data Word

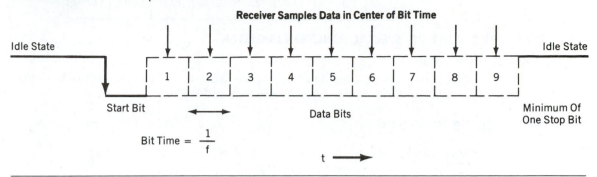

Here, as in the case for mode 0, the baud rate is much higher than standard communication rates. This high data rate is needed in many multi-processor applications. Data can be collected quickly from an extensive network of communicating microcontrollers if high baud rates are employed.

The conditions for setting RI for mode 2 are similar to mode 1: RI must be 0 before the last bit is received, *and* SM2 must be 0 *or* the ninth data bit must be a 1. Setting RI based upon the state of SM2 in the receiving 8051 and the state of bit 9 in the transmitted message makes multiprocessing possible by enabling some receivers to be interrupted by certain messages, while other receivers ignore those messages. Only those 8051's that have SM2 set to 0 will be interrupted by received data which has the ninth data bit set to 0; those with SM2 set to 1 will not be interrupted by messages with data bit 9 at 0. *All* receivers will be interrupted by data words that have the ninth data bit set to 1; the state of SM2 will not block reception of such messages.

This scheme allows the transmitting computer to "talk" to selected receiving computers without interrupting other receiving computers. Receiving computers can be commanded by the "talker" to "listen" or "deafen" by transmitting coded byte(s) with the ninth data bit set to 1. The 1 in data bit 9 interrupts all receivers, instructing those that are programmed to respond to the coded byte(s) to program the state of SM2 in their respective SCON registers. Selected listeners then respond to the bit 9 set to 0 messages, while all other receivers ignore these messages. The talker can change the mix of listeners by transmitting bit 9 set to 1 messages that instruct new listeners to set SM2 to 0, while others are instructed to set SM2 to 1.

Serial Data Mode 3

Mode 3 is identical to mode 2 except that the baud rate is determined exactly as in mode 1, using Timer 1 to generate communication frequencies.

Interrupts

A computer program has only two ways to determine the conditions that exist in internal and external circuits. One method uses software instructions that jump on the states of flags and port pins. The second responds to hardware signals, called interrupts, that force the program to call a sub-routine. Software techniques use up processor time that could be devoted to other tasks; interrupts take processor time only when action by the program is needed. Most applications of microcontrollers involve responding to events quickly enough to control the environment that generates the events (generically termed "real-

FIGURE 2.17   IE and IP Function Registers

	7	6	5	4	3	2	1	0
	EA	—	ET2	ES	ET1	EX1	ET0	EX0

THE INTERRUPT ENABLE (IE) SPECIAL FUNCTION REGISTER

Bit	Symbol	Function
7	EA	Enable interrupts bit. Cleared to 0 by program to disable all interrupts; set to 1 to permit individual interrupts to be enabled by their enable bits.
6	—	Not implemented.
5	ET2	Reserved for future use.
4	ES	Enable serial port interrupt. Set to 1 by program to enable serial port interrupt; cleared to 0 to disable serial port interrupt.
3	ET1	Enable timer 1 overflow interrupt. Set to 1 by program to enable timer 1 overflow interrupt; cleared to 0 to disable timer 1 overflow interrupt.
2	EX1	Enable external interrupt 1. Set to 1 by program to enable $\overline{INT1}$ interrupt; cleared to 0 to disable $\overline{INT1}$ interrupt.
1	ET0	Enable timer 0 overflow interrupt. Set to 1 by program to enable timer 0 overflow interrupt; cleared to 0 to disable timer 0 overflow interrupt.
0	EX0	Enable external interrupt 0. Set to 1 by program to enable $\overline{INT0}$ interrupt; cleared to 0 to disable $\overline{INT0}$ interrupt.

Bit addressable as IE.0 to IE.7

	7	6	5	4	3	2	1	0
	—	—	PT2	PS	PT1	PX1	PT0	PX0

THE INTERRUPT PRIORITY (IP) SPECIAL FUNCTION REGISTER

Bit	Symbol	Function
7	—	Not implemented.
6	—	Not implemented.
5	PT2	Reserved for future use.
4	PS	Priority of serial port interrupt. Set/cleared by program.
3	PT1	Priority of timer 1 overflow interrupt. Set/cleared by program.
2	PX1	Priority of external interrupt 1. Set/cleared by program.
1	PT0	Priority of timer 0 overflow interrupt. Set/cleared by program.
0	PX0	Priority of external interrupt 0. Set/cleared by program.

Note: Priority may be 1 (highest) or 0 (lowest)

Bit addressable as IP.0 to IP.7

time programming''). Interrupts are often the only way in which real-time programming can be done successfully.

Interrupts may be generated by internal chip operations or provided by external sources. Any interrupt can cause the 8051 to perform a hardware call to an interrupt-handling subroutine that is located at a predetermined (by the 8051 designers) absolute address in program memory.

Five interrupts are provided in the 8051. Three of these are generated automatically by internal operations: timer flag 0, timer flag 1, and the serial port interrupt (RI *or* TI). Two interrupts are triggered by external signals provided by circuitry that is connected to pins $\overline{INT0}$ and $\overline{INT1}$ (port pins P3.2 and P3.3).

All interrupt functions are under the control of the program. The programmer is able to alter control bits in the interrupt enable register (IE), the interrupt priority register (IP), and the timer control register (TCON). The program can block all or any combination of the interrupts from acting on the program by suitably setting or clearing bits in these registers. The IE and IP registers are shown in Figure 2.17.

After the interrupt has been handled by the interrupt subroutine, which is placed by the programmer at the interrupt location in program memory, the interrupted program must resume operation at the instruction where the interrupt took place. Program resumption is done by storing the interrupted PC address on the stack in RAM before changing the PC to the interrupt address in ROM. The PC address will be restored from the stack after an RETI instruction is executed at the end of the interrupt subroutine.

Timer Flag Interrupt

When a timer/counter overflows, the corresponding timer flag, TF0 or TF1, is set to 1. The flag is cleared to 0 when the resulting interrupt generates a program call to the appropriate timer subroutine in memory.

Serial Port Interrupt

If a data byte is received, an interrupt bit, RI, is set to 1 in the SCON register. When a data byte has been transmitted an interrupt bit, TI, is set in SCON. These are ORed together to provide a single interrupt to the processor: the serial port interrupt. These bits are *not* cleared when the interrupt-generated program call is made by the processor. The program that handles serial data communication *must* reset RI or TI to 0 to enable the next data communication operation.

External Interrupts

Pins $\overline{INT0}$ and $\overline{INT1}$ are used by external circuitry. Inputs on these pins can set the interrupt flags IE0 and IE1 in the TCON register to 1 by two different methods. The IEX flags may be set when the $\overline{INTX}$ pin signal reaches a low *level,* or the flags may be set when a high-to-low *transition* takes place on the $\overline{INTX}$ pin. Bits IT0 and IT1 in TCON program the $\overline{INTX}$ pins for low-level interrupt when set to 0 and program the $\overline{INTX}$ pins for transition interrupt when set to 1.

Flags IEX will be reset when a transition-generated interrupt is accepted by the processor and the interrupt subroutine is accessed. It is the responsibility of the system designer and programmer to reset *any* level-generated external interrupts when they are serviced by the program. The external circuit *must* remove the low level before an RETI is executed. Failure to remove the low will result in an immediate interrupt after RETI, from the same source.

Reset

A reset can be considered to be the ultimate interrupt because the program may not block the action of the voltage on the RST pin. This type of interrupt is often called "non-maskable," since no combination of bits in any register can stop, or mask the reset action. Unlike other interrupts, the PC is not stored for later program resumption; a reset is an absolute command to jump to program address 0000h and commence running from there.

Whenever a high level is applied to the RST pin, the 8051 enters a reset condition. After the RST pin is brought low, the internal registers will have the values shown in the following table:

REGISTER	VALUE(HEX)
PC	0000
DPTR	0000
A	00
B	00
SP	07
PSW	00
P0–3	FF
IP	XXX00000b
IE	0XX00000b
TCON	00
TMOD	00
TH0	00
TL0	00
TH1	00
TL1	00
SCON	00
SBUF	XX
PCON	0XXXXXXXb

Internal RAM is not changed by a reset; however, the states of the internal RAM when power is first applied to the 8051 are random. Register bank 0 is selected upon reset as all bits in PSW are 0.

Interrupt Control

The program must be able, at critical times, to inhibit the action of some or all of the interrupts so that crucial operations can be finished. The IE register holds the programmable bits that can enable or disable all the interrupts as a group, or if the group is enabled, each individual interrupt source can be enabled or disabled.

Often, it is desirable to be able to set priorities among competing interrupts that may conceivably occur simultaneously. The IP register bits may be set by the program to assign priorities among the various interrupt sources so that more important interrupts can be serviced first should two or more interrupts occur at the same time.

Interrupt Enable/Disable

Bits in the IE register are set to 1 if the corresponding interrupt source is to be enabled and set to 0 to disable the interrupt source. Bit EA is a master, or "global," bit that can enable or disable all of the interrupts.

Interrupt Priority

Register IP bits determine if any interrupt is to have a high or low priority. Bits set to 1 give the accompanying interrupt a high priority while a 0 assigns a low priority. Interrupts with a high priority can interrupt another interrupt with a lower priority; the low priority interrupt continues after the higher is finished.

If two interrupts with the same priority occur at the same time, then they have the following ranking:

1. IE0
2. TF0
3. IE1
4. TF1
5. Serial = RI OR TI

The serial interrupt could be given the highest priority by setting the PS bit in IP to 1, and all others to 0.

Interrupt Destinations

Each interrupt source causes the program to do a hardware call to one of the dedicated addresses in program memory. It is the responsibility of the programmer to place a routine at the address that will service the interrupt.

The interrupt saves the PC of the program, which is running at the time the interrupt is serviced on the stack in internal RAM. A call is then done to the appropriate memory location. These locations are shown in the following table:

INTERRUPT	ADDRESS(HEX)
IE0	0003
TF0	000B
IE1	0013
TF1	001B
SERIAL	0023

A RETI instruction at the end of the routine restores the PC to its place in the interrupted program and resets the interrupt logic so that another interrupt can be serviced. Interrupts that occur but are ignored due to any blocking condition (IE bit not set or a higher priority interrupt already in process) *must* persist until they are serviced, or they will be *lost*. This requirement applies primarily to the level-activated $\overline{INTX}$ interrupts.

Software Generated Interrupts

When *any* interrupt flag is set to 1 *by any means*, an interrupt is generated unless blocked. This means that the program itself can cause interrupts of any kind to be generated simply by setting the desired interrupt flag to 1 using a program instruction.

Summary

The internal hardware configuration of the 8051 registers and control circuits have been examined at the functional block diagram level. The 8051 may be considered to be a collection of RAM, ROM, and addressable registers that have some unique functions.

SPECIAL-FUNCTION REGISTERS

Register	Bit	Primary Function	Bit Addressable
A	8	Math, data manipulation	Y
B	8	Math	Y
PC	16	Addressing program bytes	N
DPTR	16	Addressing code and external data	N
SP	8	Addressing internal RAM stack data	N
PSW	8	Processor status	Y
P0–P3	8	Store I/O port data	Y
TH0/TL0	8/8	Timer/counter 0	N
TH1/TL1	8/8	Timer/counter 1	N
TCON	8	Timer/counter control	Y
TMOD	8	Timer/counter control	N
SBUF	8	Serial port data	N
SCON	8	Serial port control	Y
PCON	8	Serial port control, user flags	N
IE	8	Interrupt enable control	Y
IP	8	Interrupt priority control	Y

DATA AND PROGRAM MEMORY

Internal	Bytes	Function
RAM	128	R0–R7 registers, data storage, stack
ROM	4K	Program storage

External	Bytes	Function
RAM	64K	Data storage
ROM	64K	Program storage

EXTERNAL CONNECTION PINS

		Function
Port pins	36	I/O, external memory, interrupts
Oscillator	2	Clock
Power	2	

Questions

Find the following using the information provided in Chapter 2.

1. Size of the internal RAM.

2. Internal ROM size in the 8031.

3. Execution time of a single cycle instruction for a 6 megahertz crystal.

4. The 16-bit data addressing registers and their functions.

5. Registers that can do division.

6. The flags that are stored in the PSW.

7. Which register holds the serial data interrupt bits TI and RI.

8. Address of the stack when the 8051 is reset.

9. Number of register banks and their addresses.

10. Ports used for external memory access.

11. The bits that determine timer modes and the register that holds these bits.

12. Address of a subroutine that handles a timer 1 interrupt.

13. Why a low-address byte latch for external memory is needed.

14. How an I/O pin can be both an input and output.

15. Which port has no alternate functions.

16. The maximum pulse rate that can be counted on pin T1 if the oscillator frequency is 6 megahertz.

17. Which bits in which registers must be set to give the serial data interrupt the highest priority.

18. The baud rate for the serial port in mode 0 for a 6 megahertz crystal.

19. The largest possible time delay for a timer in mode 1 if a 6 megahertz crystal is used.

20. The setting of TH1, in timer mode 2, to generate a baud rate of 1200 if the serial port is in mode 1 and an 11.059 megahertz crystal is in use. Find the setting for both values of SMOD.

21. The address of the PCON special-function register.

22. The time it will take a timer in mode 1 to overflow if initially set to 03AEh with a 6 megahertz crystal.

23. Which bits in which registers must be set to 1 to have timer 0 count input pulses on pin T0 in timer mode 0.

24. The register containing GF0 and GF1.

25. The signal that reads external ROM.

26. When used in multiprocessing, which bit in which register is used by a transmitting 8051 to signal receiving 8051's that an interrupt should be generated.

27. The two conditions under which program opcodes are fetched from external, rather than internal, memory.

28. Which bits in which register(s) must be set to make $\overline{INT0}$ level activated, and $\overline{INT1}$ edge triggered.

29. The address of the interrupt program for the $\overline{INT0}$ level-generated interrupt.

30. The bit address of bit 4 of RAM byte 2Ah.

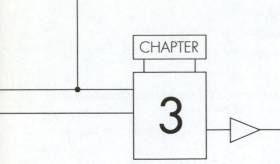

CHAPTER

3

Moving Data

Chapter Outline

Introduction
Addressing Modes
External Data Moves
PUSH and POP Opcodes

Data Exchanges
Example Programs
Summary

Introduction

A computer typically spends more time moving data from one location to another than it spends on any other operation. It is not surprising, therefore, to find that more instructions are provided for moving data than for any other type of operation.

Data is stored at a *source* address and moved (actually, the data is *copied*) to a *destination* address. The ways by which these addresses are specified are called the *addressing modes*. The 8051 mnemonics are written with the *destination* address named *first,* followed by the source address.

A detailed study of the operational codes (opcodes) of the 8051 begins in this chapter. Although there are 28 distinct mnemonics that copy data from a source to a destination, they may be divided into the following three main types:

1. MOV destination, source
2. PUSH source or POP destination
3. XCH destination, source

The following four addressing modes are used to access data:

1. Immediate addressing mode
2. Register addressing mode

3. Direct addressing mode

4. Indirect addressing mode

The MOV opcodes involve data transfers within the 8051 memory. This memory is divided into the following four distinct physical parts:

1. Internal RAM

2. Internal special-function registers

3. External RAM

4. Internal and external ROM

Finally, the following five types of opcodes are used to move data:

1. MOV

2. MOVX

3. MOVC

4. PUSH and POP

5. XCH

Addressing Modes

The way in which the data sources or destination addresses are specified in the mnemonic that moves that data determines the addressing mode. Figure 3.1 diagrams the four addressing modes: immediate, register, direct, and indirect.

Immediate Addressing Mode

The simplest way to get data to a destination is to make the source of the data part of the opcode. The data source is then immediately available as part of the instruction itself.

When the 8051 executes an immediate data move, the program counter is automatically incremented to point to the byte(s) following the opcode byte in the program memory. Whatever data is found there is copied to the destination address.

The mnemonic for immediate data is the pound sign (#). Occasionally, in the rush to meet a deadline, one forgets to use the # for immediate data. The resulting opcode is often a legal command that is assembled with no objections by the assembler. This omission guarantees that the rush will continue.

Register Addressing Mode

Certain register names may be used as part of the opcode mnemonic as sources or destinations of data. Registers A, DPTR, and R0 to R7 may be named as part of the opcode mnemonic. Other registers in the 8051 may be addressed using the direct addressing mode. Some assemblers can equate many of the direct addresses to the register name (as is the case with the assembler discussed in this text) so that register names may be used in lieu of register addresses. Remember that the registers used in the opcode as R0 to R7 are the ones that are *currently* chosen by the bank-select bits, RS0 and RS1 in the PSW.

The following table shows all possible MOV opcodes using immediate and register addressing modes:

Mnemonic	Operation
MOV A,#n	Copy the immediate data byte n to the A register
MOV A,Rr	Copy data from register Rr to register A
MOV Rr,A	Copy data from register A to register Rr
MOV Rr,#n	Copy the immediate data byte n to register Rr
MOV DPTR,#nn	Copy the immediate 16-bit number nn to the DPTR register

FIGURE 3.1 Addressing Modes

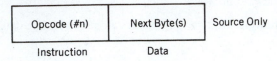

Immediate Addressing Mode

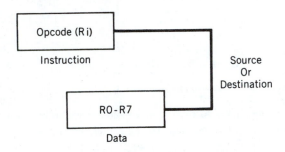

Register Addressing Mode

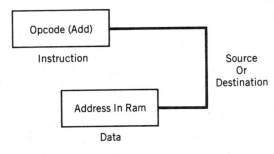

Direct Addressing Mode

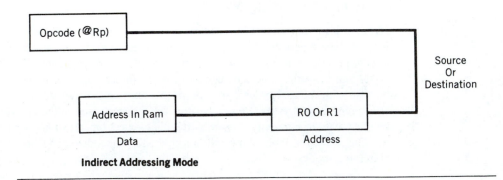

Indirect Addressing Mode

A data MOV does not alter the contents of the data source address. A *copy* of the data is made from the source and moved to the destination address. The contents of the destination address are replaced by the source address contents. The following table shows examples of MOV opcodes with immediate and register addressing modes:

Mnemonic	Operation
MOV A,#0F1h	Move the immediate data byte F1h to the A register
MOV A,R0	Copy the data in register R0 to register A
MOV DPTR,#0ABCDh	Move the immediate data bytes ABCDh to the DPTR
MOV R5,A	Copy the data in register A to register R5
MOV R3,#1Ch	Move the immediate data byte 1Ch to register R3

 CAUTION ————————————————————————————————————

It is impossible to have immediate data as a destination.

All numbers *must* start with a decimal number (0–9), or the assembler assumes the number is a *label*.

Register-to-register moves using the register addressing mode occur between registers A and R0 to R7.

Direct Addressing Mode

All 128 bytes of internal RAM and the SFRs may be addressed directly using the single-byte address assigned to each RAM location and each special-function register.

Internal RAM uses addresses from 00 to 7Fh to address each byte. The SFR addresses exist from 80h to FFh at the following locations:

SFR	ADDRESS (HEX)
A	0E0
B	0F0
DPL	82
DPH	83
IE	0A8
IP	0B8
P0	80
P1	90
P2	0A0
P3	0B0
PCON	87
PSW	0D0
SBUF	99
SCON	98
SP	81
TCON	88
TMOD	89
TH0	8C
TL0	8A
TH1	8D
TL1	8B

Note the use of a leading 0 for all numbers that begin with an alphabetic (alpha) character.

RAM addresses 00 to 1Fh are *also* the locations assigned to the four banks of eight working registers, R0 to R7. This assignment means that R2 of register bank 0 can be addressed in the register mode as R2 or in the direct mode as 02h. The direct addresses of the working registers are as follows:

BANK REGISTER		ADDRESS (HEX)	BANK REGISTER		ADDRESS (HEX)
0	R0	00	2	R0	10
0	R1	01	2	R1	11
0	R2	02	2	R2	12
0	R3	03	2	R3	13
0	R4	04	2	R4	14
0	R5	05	2	R5	15
0	R6	06	2	R6	16
0	R7	07	2	R7	17
1	R0	08	3	R0	18
1	R1	09	3	R1	19
1	R2	0A	3	R2	1A
1	R3	0B	3	R3	1B
1	R4	0C	3	R4	1C
1	R5	0D	3	R5	1D
1	R6	0E	3	R6	1E
1	R7	0F	3	R7	1F

Only one bank of working registers is *active* at any given time. The PSW special-function register holds the bank-select bits, RS0 and RS1, which determine which register bank is in use.

When the 8051 is reset, RS0 and RS1 are set to 00b to select the working registers in bank 0, located from 00h to 07h in internal RAM. Reset also sets SP to 07h, and the stack will grow *up* as it is used. This growing stack will overwrite the register banks above bank 0. Be *sure* to set the SP to a number above those of any working registers the program may use.

The programmer may choose any other bank by setting RS0 and RS1 as desired; this bank change is often done to "save" one bank and choose another when servicing an interrupt or using a subroutine.

The moves made possible using direct, immediate, and register addressing modes are as follows:

Mnemonic	**Operation**
MOV A,add	Copy data from direct address add to register A
MOV add,A	Copy data from register A to direct address add
MOV Rr,add	Copy data from direct address add to register Rr
MOV add,Rr	Copy data from register Rr to direct address add
MOV add,#n	Copy immediate data byte n to direct address add
MOV add1,add2	Copy data from direct address add2 to direct address add1

The following table shows examples of MOV opcodes using direct, immediate, and register addressing modes:

Mnemonic	**Operation**
MOV A,80h	Copy data from the port 0 pins to register A
MOV 80h,A	Copy data from register A to the port 0 latch
MOV 3Ah,#3Ah	Copy immediate data byte 3Ah to RAM location 3Ah
MOV R0,12h	Copy data from RAM location 12h to register R0

MOV 8Ch,R7	Copy data from register R7 to timer 0 high byte
MOV 5Ch,A	Copy data from register A to RAM location 5Ch
MOV 0A8h,77h	Copy data from RAM location 77h to IE register

 CAUTION

MOV instructions that refer to direct addresses above 7Fh that are not SFRs will result in errors. The SFRs are physically on the chip; all other addresses above 7Fh do not physically exist.

Moving data to a port changes the port *latch;* moving data from a port gets data from the port *pins.*

Moving data from a direct address to itself is not predictable and could lead to errors.

Indirect Addressing Mode

For all the addressing modes covered to this point, the source or destination of the data is an absolute number or a name. Inspection of the opcode reveals exactly what are the addresses of the destination and source. For example, the opcode MOV A,R7 says that the A register will get a copy of whatever data is in register R7; MOV 33h,#32h moves the hex number 32 to hex RAM address 33.

The indirect addressing mode uses a register to *hold* the actual address that will finally be used in the data move; the register itself is *not* the address, but rather the number *in* the register. Indirect addressing for MOV opcodes uses register R0 or R1, often called "data pointers," to hold the address of one of the data locations in RAM from address 00h to 7Fh. The number that is in the pointing register (Rp) cannot be known unless the history of the register is known. The mnemonic symbol used for indirect addressing is the "at" sign, which is printed as @.

The moves made possible using immediate, direct, register and indirect addressing modes are as follows:

Mnemonic	**Operation**
MOV @Rp,#n	Copy the immediate byte n to the address in Rp
MOV @Rp,add	Copy the contents of add to the address in Rp
MOV @Rp,A	Copy the data in A to the address in Rp
MOV add,@Rp	Copy the contents of the address in Rp to add
MOV A,@Rp	Copy the contents of the address in Rp to A

The following table shows examples of MOV opcodes, using immediate, register, direct, and indirect modes

Mnemonic	**Operation**
MOV A,@R0	Copy the contents of the address in R0 to the A register
MOV @R1,#35h	Copy the number 35h to the address in R1
MOV add,@R0	Copy the contents of the address in R0 to add
MOV @R1,A	Copy the contents of A to the address in R1
MOV @R0,80h	Copy the contents of the port 0 pins to the address in R0

 CAUTION

The number in register Rp must be a RAM address.

Only registers R0 or R1 may be used for indirect addressing.

FIGURE 3.2   External Addressing using MOVX and MOVC

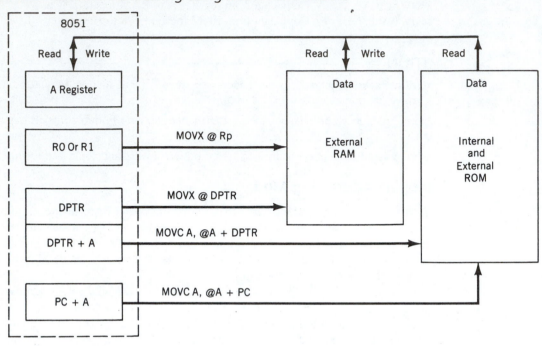

External Data Moves

As discussed in Chapter 2, it is possible to expand RAM and ROM memory space by adding external memory chips to the 8051 microcontroller. The external memory can be as large as 64K bytes for each of the RAM and ROM memory areas. Opcodes that access this external memory *always* use indirect addressing to specify the external memory.

Figure 3.2 shows that registers R0, R1, and the aptly named DPTR can be used to hold the address of the data byte in external RAM. R0 and R1 are limited to external RAM address ranges of 00h to 0FFh, while the DPTR register can address the maximum RAM space of 0000h to 0FFFFh.

An X is added to the MOV mnemonics to serve as a reminder that the data move is external to the 8051, as shown in the following table.

Mnemonic	Operation
MOVX A,@Rp	Copy the contents of the external address in Rp to A
MOVX A,@DPTR	Copy the contents of the external address in DPTR to A
MOVX @Rp,A	Copy data from A to the external address in Rp
MOVX @DPTR,A	Copy data from A to the external address in DPTR

The following table shows examples of external moves using register and indirect addressing modes:

Mnemonic	Operation
MOVX @DPTR,A	Copy data from A to the 16-bit address in DPTR
MOVX @R0,A	Copy data from A to the 8-bit address in R0

MOVX A,@R1	Copy data from the 8-bit address in R1 to A
MOVX A,@DPTR	Copy data from the 16-bit address in DPTR to A

 **CAUTION**

All external data moves must involve the A register.

Rp can address 256 bytes; DPTR can address 64K bytes.

MOVX is normally used with external RAM or I/O addresses.

Note that there are two sets of RAM addresses between 00 and 0FFh: one internal and one external to the 8051.

Code Memory Read-Only Data Moves

Data moves between RAM locations and 8051 registers are made by using MOV and MOVX opcodes. The data is usually of a temporary or "scratch pad" nature and disappears when the system is powered down.

There are times when access to a preprogrammed mass of data is needed, such as when using tables of predefined bytes. This data must be permanent to be of repeated use and is stored in the program ROM using assembler directives that store programmed data anywhere in ROM that the programmer wishes.

Access to this data is made possible by using indirect addressing and the A register in conjunction with either the PC or the DPTR, as shown in Figure 3.2. In both cases, the number in register A is *added* to the pointing register to form the address in ROM where the desired data is to be found. The data is then fetched from the ROM address so formed and placed in the A register. The original data in A is lost, and the addressed data takes its place.

As shown in the following table, the letter C is added to the MOV mnemonic to highlight the use of the opcodes for moving data from the source address in the Code ROM to the A register in the 8051:

Mnemonic	Operation
MOVC A,@A+DPTR	Copy the code byte, found at the ROM address formed by adding A and the DPTR, to A
MOVC A,@A+PC	Copy the code byte, found at the ROM address formed by adding A and the PC, to A

Note that the DPTR and the PC are not changed; the A register contains the ROM byte found at the address formed.

The following table shows examples of code ROM moves using register and indirect addressing modes:

Mnemonic	Operation
MOV DPTR,#1234h	Copy the immediate number 1234h to the DPTR
MOV A,#56h	Copy the immediate number 56h to A
MOVC A,@A+DPTR	Copy the contents of address 128Ah to A
MOVC A,@A+PC	Copies the contents of address 4059h to A if the PC contained 4000h and A contained 58h when the opcode is executed.

—▷— CAUTION 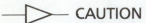 ————————————————————————————

> The PC is incremented by one (to point to the next instruction) *before* it is added to A to form the final address of the code byte.
>
> All data is moved *from* the code memory *to* the A register.
>
> MOVC is normally used with internal or external ROM and can address 4K of internal or 64K bytes of external code.

PUSH and POP Opcodes

The PUSH and POP opcodes specify the direct address of the data. The data moves between an area of internal RAM, known as the stack, and the specified direct address. The stack pointer special-function register (SP) contains the address in RAM where data *from* the source address will be PUSHed, or where data to be POPed *to* the destination address is found. The SP register actually is used in the indirect addressing mode but is *not* named in the mnemonic. It is *implied* that the SP holds the indirect address whenever PUSHing or POPing. Figure 3.3 shows the operation of the stack pointer as data is PUSHed or POPed to the stack area in internal RAM.

A PUSH opcode copies data from the source address to the stack. SP is *incremented* by one *before* the data is copied to the internal RAM location contained in SP so that the data is stored from low addresses to high addresses in the internal RAM. The stack grows *up* in memory as it is PUSHed. Excessive PUSHing can make the stack exceed 7Fh (the top of internal RAM), after which point data is lost.

A POP opcode copies data from the stack to the destination address. SP is *decremented* by one *after* data is copied from the stack RAM address to the direct destination to ensure that data placed on the stack is retrieved in the same order as it was stored.

The PUSH and POP opcodes behave as explained in the following table:

Mnemonic	Operation
PUSH add	Increment SP; copy the data in add to the internal RAM address contained in SP
POP add	Copy the data from the internal RAM address contained in SP to add; decrement the SP

FIGURE 3.3   PUSH and POP the Stack

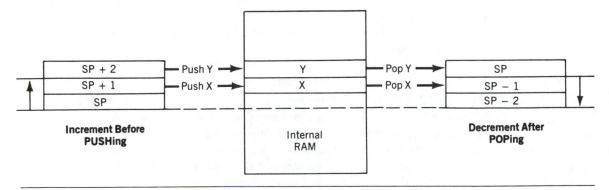

The SP register is set to 07h when the 8051 is reset, which is the same direct address in internal RAM as register R7 in bank 0. The first PUSH opcode would write data to R0 of bank 1. The SP should be initialized by the programmer to point to an internal RAM address above the highest address likely to be used by the program.

The following table shows examples of PUSH and POP opcodes:

Mnemonic	Operation
MOV 81h,#30h	Copy the immediate data 30h to the SP
MOV R0, #0ACh	Copy the immediate data ACh to R0
PUSH 00h	SP = 31h; address 31h contains the number ACh
PUSH 00h	SP = 32h; address 32h contains the number ACh
POP 01h	SP = 31h; register R1 now contains the number ACh
POP 80h	SP = 30h; port 0 latch now contains the number ACh

CAUTION

When the SP reaches FFh it "rolls over" to 00h (R0).

RAM ends at address 7Fh, PUSHes above 7Fh result in errors.

The SP is usually set at addresses above the register banks.

The SP may be PUSHed and POPed to the stack.

Note that direct addresses, *not* register names, must be used for most registers. The stack mnemonics have no way of knowing which bank is in use.

Data Exchanges

MOV, PUSH, and POP opcodes all involve copying the data found in the source address to the destination address; the original data in the source is not changed. Exchange instructions actually move data in two directions: from source to destination and from destination to source. All addressing modes except immediate may be used in the XCH (exchange) opcodes:

Mnemonic	Operation
XCH A,Rr	Exchange data bytes between register Rr and A
XCH A,add	Exchange data bytes between add and A
XCH A,@Rp	Exchange data bytes between A and address in Rp
XCHD A,@Rp	Exchange lower nibble between A and address in Rp

Exchanges between A and any port location copy the data on the port *pins* to A, while the data in A is copied to the port *latch*. Register A is used for so many instructions that the XCH opcode provides a very convenient way to "save" the contents of A without the necessity of using a PUSH opcode and then a POP opcode.

The following table shows examples of data moves using exchange opcodes:

Mnemonic	Operation
XCH A,R7	Exchange bytes between register A and register R7
XCH A,0F0h	Exchange bytes between register A and register B
XCH A,@R1	Exchange bytes between register A and address in R1
XCHD A,@R1	Exchange lower nibble in A and the address in R1

 CAUTION

All exchanges are *internal* to the 8051.

All exchanges use register A.

When using XCHD, the upper nibble of A and the upper nibble of the address location in Rp do not change.

This section concludes the listing of the various data moving instructions; the remaining sections will concentrate on using these opcodes to write short programs.

Example Programs

Programming is at once a skill and an art. Just as anyone may learn to play a musical instrument after sufficient instruction and practice, so may anyone learn to program a computer. Some individuals, however, have a gift for programming that sets them apart from their peers with the same level of experience, just as some musicians are more talented than their contemporaries.

Gifted or not, you will not become adept at programming until you have written and rewritten many programs. The emphasis here is on practice; you can read many books on how to ride a bicycle, but you do not know how to ride until you do it.

If some of the examples and problems seem trivial or without any "real-world" application, remember the playing of scales on a piano by a budding musician. Each example will be done using several methods; the best method depends upon what resource is in short supply. If programming time is valuable, then the best program is the one that uses the fewest lines of code; if either ROM or execution time is limited, then the program that uses the fewest code bytes is best.

EXAMPLE PROBLEM 3.1

Copy the byte in TCON to register R2 using at least four different methods.

■ **Method 1:** Use the direct address for TCON (88h) and register R2.

Mnemonic	Operation
MOV R2,88h	Copy TCON to R2

■ **Method 2:** Use the direct addresses for TCON and R2.

Mnemonic	Operation
MOV 02h,88h	Copy TCON to direct address 02h (R2)

■ **Method 3:** Use R1 as a pointer to R2 and use the address of TCON.

Mnemonic	Operation
MOV R1,#02h	Use R1 as a pointer to R2
MOV @R1,88h	Copy TCON byte to address in R1 (02h = R2)

■ **Method 4:** Push the contents of TCON into direct address 02h (R2).

Mnemonic	Operation
MOV 81h,#01h	Set the SP to address 01h in RAM
PUSH 88h	Push TCON (88h) to address 02h (R2)

EXAMPLE PROBLEM 3.2

Set timer T0 to an initial setting of 1234h.

Use the direct address with an immediate number to set TH0 and TL0.

Mnemonic	**Operation**
MOV 8Ch,#12h	Set TH0 to 12h
MOV 8Ah,#34h	Set TL0 to 34h
	Totals: 6 bytes, 2 lines

EXAMPLE PROBLEM 3.3

Put the number 34h in registers R5, R6, and R7.

■ **Method 1:** Use an immediate number and register addressing.

Mnemonic	**Operation**
MOV R5,#34h	Copy 34h to R5
MOV R6,#34h	Copy 34h to R6
MOV R7,#34h	Copy 34h to R7
	Totals: 6 bytes, 3 lines

■ **Method 2:** Since the number is the same for each register, put the number in A and MOV A to each register.

Mnemonic	**Operation**
MOV A,#34h	Copy a 34h to A
MOV R5,A	Copy A to R5
MOV R6,A	Copy A to R6
MOV R7,A	Copy A to R7
	Totals: 5 bytes, 4 lines

■ **Method 3:** Copy one direct address to another.

Mnemonic	**Operation**
MOV R5,#34h	Copy 34h to register R5
MOV 06h,05h	Copy R5 (add 05) to R6 (add 06)
MOV 07h,06h	Copy R6 to R7
	Totals: 8 bytes, 3 lines

EXAMPLE PROBLEM 3.4

Put the number 8Dh in RAM locations 30h to 34h.

■ **Method 1:** Use the immediate number to a direct address:

Mnemonic	Operation
MOV 30h,#8Dh	Copy the number 8Dh to RAM address 30h
MOV 31h,#8Dh	Copy the number 8Dh to RAM address 31h
MOV 32h,#8Dh	Copy the number 8Dh to RAM address 32h
MOV 33h,#8Dh	Copy the number 8Dh to RAM address 33h
MOV 34h,#8Dh	Copy the number 8Dh to RAM address 34h
	Totals: 15 bytes, 5 lines

■ **Method 2:** Using the immediate number in each instruction uses bytes; use a register to hold the number:

Mnemonic	Operation
MOV A,#8Dh	Copy the number 8Dh to the A register
MOV 30h,A	Copy the contents of A to RAM location 30h
MOV 31h,A	Copy the contents of A to the remaining addresses
MOV 32h,A	
MOV 33h,A	
MOV 34h,A	Totals: 12 bytes, 6 lines

■ **Method 3:** There must be a way to avoid naming each address; the PUSH opcode can increment to each address:

Mnemonic	Operation
MOV 30h,#8Dh	Copy the number 8Dh to RAM address 30h
MOV 81h,#30h	Set the SP to 30h
PUSH 30h	Push the contents of 30h (=8Dh) to address 31h
PUSH 30h	Continue pushing to address 34h
PUSH 30h	
PUSH 30h	Totals: 14 bytes, 6 lines

 COMMENT

Indirect addressing with the number in A and the indirect address in R1 could be done; however, R1 would have to be loaded with each address from 30h to 34h. Loading R1 would take a total of 17 bytes and 11 lines of code. Indirect addressing is advantageous when we have opcodes that can change the contents of the pointing registers automatically.

Summary

The opcodes that move data between locations within the 8051 and between the 8051 and external memory have been discussed. The general form and results of these instructions are as follows.

Instruction Type	Result
MOV destination,source	Copy data from the internal RAM source address to the internal RAM destination address

MOVC A,source	Copy internal or external program memory byte from the source to register A
MOVX destination,source	Copy byte to or from external RAM to register A
PUSH source	Copy byte to internal RAM stack from internal RAM source
POP destination	Copy byte from internal RAM stack to internal RAM destination
XCH A,source	Exchange data between register A and the internal RAM source
XCHD A,source	Exchange lower nibble between register A and the internal RAM source

There are four addressing modes: an immediate number, a register name, a direct internal RAM address, and an indirect address contained in a register.

Problems

Write programs that will accomplish the desired tasks listed below, using as few lines of code as possible. Use only opcodes that have been covered up to this chapter. Comment on each line of code.

1. Place the number 3Bh in internal RAM locations 30h to 32h.

2. Copy the data at internal RAM location F0h to R0 and R3.

3. Set the SP at the byte address just above the last working register address.

4. Exchange the contents of the SP and the PSW.

5. Copy the byte at internal RAM address 27h to external RAM address 27h.

6. Set Timer 1 to A23Dh.

7. Copy the contents of DPTR to registers R0 (DPL) and R1 (DPH).

8. Copy the data in external RAM location 0123h to TL0 and the data in external RAM location 0234h to TH0.

9. Copy the data in internal RAM locations 12h to 15h to internal RAM locations 20h to 23h: Copy 12h to 20h, 13h to 21h, etc.

10. Set the SP register to 07h and PUSH the SP register on the stack; predict what number is PUSHed to address 08h.

11. Exchange the contents of the B register and external RAM address 02CFh.

12. Rotate the bytes in registers R0 to R3; copy the data in R0 to R1, R1 to R2, R2 to R3, and R3 to R0.

13. Copy the external code byte at address 007Dh to the SP.

14. Copy the data in register R5 to external RAM address 032Fh.

15. Copy the internal code byte at address 0300h to external RAM address 0300h.

16. Swap the bytes in timer 0; put TL0 in TH0 and TH0 in TL0.

17. Store DPTR in external RAM locations 0123h (DPL) and 02BCh (DPH).

18. Exchange both low nibbles of registers R0 and R1; put the low nibble of R0 in R1, and the low nibble of R1 in R0.

19. Store the contents of register R3 at the internal RAM address contained in R2. (Be sure the address in R2 is legal.)

20. Store the contents of RAM location 20h at the address contained in RAM location 08h.

21. Store register A at the internal RAM location address in register A.

22. Copy program bytes 0100h to 0102h to internal RAM locations 20h to 22h.

23. Copy the data on the pins of port 2 to the port 2 latch.

24. PUSH the contents of the B register to TMOD.

25. Copy the contents of external code memory address 0040h to IE.

26. Show that a set of XCH instructions executes faster than a PUSH and POP when saving the contents of the A register.

CHAPTER

4

Logical Operations

Chapter Outline

Introduction
Byte-Level Logical Operations
Bit-Level Logical Operations

Rotate and Swap Operations
Example Programs
Summary

Introduction

One application area the 8051 is designed to fill is that of machine control. A large part of machine control concerns sensing the on–off states of external switches, making decisions based on the switch states, and then turning external circuits on or off.

Single point sensing and control implies a need for *byte* and *bit* opcodes that operate on data using Boolean operators. All 8051 RAM areas, both data and SFRs, may be manipulated using byte opcodes. Many of the SFRs, and a unique internal RAM area that is bit addressable, may be operated upon at the *individual* bit level. Bit operators are notably efficient when speed of response is needed. Bit operators yield compact program code that enhances program execution speed.

The two data levels, byte or bit, at which the Boolean instructions operate are shown in the following table:

BOOLEAN OPERATOR	8051 MNEMONIC
AND	ANL (AND logical)
OR	ORL (OR logical)
XOR	XRL (exclusive OR logical)
NOT	CPL (complement)

There are also rotate opcodes that operate only on a byte, or a byte and the carry flag, to permit limited 8- and 9-bit shift-register operations. The following table shows the rotate opcodes:

Mnemonic	Operation
RL	Rotate a byte to the left; the Most Significant Bit (MSB) becomes the Least Significant Bit (LSB)
RLC	Rotate a byte and the carry bit left; the carry becomes the LSB, the MSB becomes the carry
RR	Rotate a byte to the right; the LSB becomes the MSB
RRC	Rotate a byte and the carry to the right; the LSB becomes the carry, and the carry the MSB
SWAP	Exchange the low and high nibbles in a byte

Byte-Level Logical Operations

The byte-level logical operations use all four addressing modes for the source of a data byte. The A register or a direct address in internal RAM is the destination of the logical operation result.

Keep in mind that all such operations are done using each individual bit of the destination and source bytes. These operations, called *byte-level Boolean operations* because the entire byte is affected, are listed in the following table:

Mnemonic	Operation
ANL A,#n	AND each bit of A with the same bit of immediate number n; put the results in A
ANL A,add	AND each bit of A with the same bit of the direct RAM address; put the results in A
ANL A,Rr	AND each bit of A with the same bit of register Rr; put the results in A
ANL A,@Rp	AND each bit of A with the same bit of the contents of the RAM address contained in Rp; put the results in A
ANL add,A	AND each bit of A with the direct RAM address; put the results in the direct RAM address
ANL add,#n	AND each bit of the RAM address with the same bit in the number n; put the result in the RAM address
ORL A,#n	OR each bit of A with the same bit of n; put the results in A
ORL A,add	OR each bit of A with the same bit of the direct RAM address; put the results in A
ORL A,Rr	OR each bit of A with the same bit of register Rr; put the results in A
ORL A,@Rp	OR each bit of A with the same bit of the contents of the RAM address contained in Rp; put the results in A
ORL add,A	OR each bit of A with the direct RAM address; put the results in the direct RAM address
ORL add,#n	OR each bit of the RAM address with the same bit in the number n; put the result in the RAM address
XRL A,#n	XOR each bit of A with the same bit of n; put the results in A
XRL A,add	XOR each bit of A with the same bit of the direct RAM address; put the results in A
XRL A,Rr	XOR each bit of A with the same bit of register Rr; put the results in A
XRL A,@Rp	XOR each bit of A with the same bit of the contents of the RAM address contained in Rp; put the results in A
XRL add,A	XOR each bit of A with the direct RAM address; put the results in the direct RAM address

XRL add,#n	XOR each bit of the RAM address with the same bit in the number n; put the result in the RAM address
CLR A	Clear each bit of the A register to zero
CPL A	Complement each bit of A; every 1 becomes a 0, and each 0 becomes a 1

Note that no flags are affected unless the direct RAM address is the PSW.

Many of these byte-level operations use a direct address, which can include the port SFR addresses, as a destination. The normal source of data from a port is the port pins; the normal destination for port data is the port latch. When the *destination* of a logical operation is the direct address of a port, the *latch* register, *not* the pins, is used *both* as the *source* for the original data and then the *destination* for the altered byte of data. *Any* port operation that must first *read* the source data, logically operate on it, and then *write* it back to the source (now the destination) must use the *latch*. Logical operations that use the port as a source, but *not* as a destination, use the *pins* of the port as the source of the data.

For example, the port 0 latch contains FFh, but the pins are all driving transistor bases and are close to ground level. The logical operation

ANL P0,#0Fh

which is designed to turn the upper nibble transistors off, reads FFh from the latch, ANDs it with 0Fh to produce 0Fh as a result, and then writes it back to the latch to turn these transistors off. Reading the pins produces the result 00h, turning all transistors off, in error. But, the operation

ANL A,P0

produces A = 00h by using the port 0 pin data, which is 00h.

The following table shows byte-level logical operation examples:

Mnemonic	**Operation**
MOV A,#0FFh	A = FFh
MOV R0,#77h	R0 = 77h
ANL A,R0	A = 77h
MOV 15h,A	15h = 77h
CPL A	A = 88h
ORL 15h,#88h	15h = FFh
XRL A,15h	A = 77h
XRL A,R0	A = 00h
ANL A,15h	A = 00h
ORL A,R0	A = 77h
CLR A	A = 00h
XRL 15h,A	15h = FFh
XRL A,R0	A = 77h

Note that instructions that can use the SFR port latches as destinations are ANL, ORL, and XRL.

── ▷ ── CAUTION ──────────────────────

If the direct address destination is one of the port SFRs, the data latched in the SFR, not the pin data, is used.

No flags are affected unless the direct address is the PSW.

Only internal RAM or SFRs may be logically manipulated.

Bit-Level Logical Operations

Certain internal RAM and SFRs can be addressed by their byte addresses or by the address of each bit within a byte. Bit addressing is *very* convenient when you wish to alter a single bit of a byte, in a control register for instance, without having to wonder what you need to do to avoid altering some other crucial bit of the same byte. The assembler can also equate bit addresses to labels that make the program more readable. For example, bit 4 of TCON can become TR0, a label for the timer 0 run bit.

The ability to operate on individual bits creates the need for an area of RAM that contains data addresses that hold a single bit. Internal RAM byte addresses 20h to 2Fh serve this need and are both byte and bit addressable. The bit addresses are numbered from 00h to 7Fh to represent the 128d bit addresses (16d bytes × 8 bits) that exist from byte addresses 20h to 2Fh. Bit 0 of *byte* address 20h is *bit* address 00h, and bit 7 of *byte* address 2Fh is *bit* address 7Fh. You must know your bits from your bytes to take advantage of this RAM area.

Internal RAM Bit Addresses

The availability of individual bit addresses in internal RAM makes the use of the RAM very efficient when storing bit information. Whole bytes do not have to be used up to store one or two bits of data.

The correspondence between byte and bit addresses are shown in the following table:

BYTE ADDRESS (HEX)	BIT ADDRESSES (HEX)
20	00–07
21	08–0F
22	10–17
23	18–1F
24	20–27
25	28–2F
26	30–37
27	38–3F
28	40–47
29	48–4F
2A	50–57
2B	58–5F
2C	60–67
2D	68–6F
2E	70–77
2F	78–7F

Interpolation of this table shows, for example, the address of bit 3 of internal RAM byte address 2Ch is 63h, the bit address of bit 5 of RAM address 21h is 0Dh, and bit address 47h is bit 7 of RAM byte address 28h.

SFR Bit Addresses

All SFRs may be addressed at the byte level by using the direct address assigned to it, but not all of the SFRs are addressable at the bit level. The SFRs that are also bit addressable form the bit address by using the five most significant bits of the direct address for that SFR, together with the three least significant bits that identify the bit position from position 0 (LSB) to 7 (MSB).

The bit-addressable SFR and the corresponding bit addresses are as follows:

SFR	DIRECT ADDRESS (HEX)	BIT ADDRESSES (HEX)
A	0E0	0E0−0E7
B	0F0	0F0−0F7
IE	0A8	0A8−0AF
IP	0B8	0B8−0BF
P0	80	80−87
P1	90	90−97
P2	0A0	0A0−0A7
P3	0B0	0B0−0B7
PSW	0D0	0D0−0D7
TCON	88	88−8F
SCON	98	98−9F

The patterns in this table show the direct addresses assigned to the SFR bytes all have bits 0−3 equal to zero so that the address of the byte is also the address of the LSB. For example, bit 0E3h is bit 3 of the A register. The carry flag, which is bit 7 of the PSW, is bit addressable as 0D7h. The assembler can also "understand" more descriptive mnemonics, such as P0.5 for bit 5 of port 0, which is more formally addressed as 85h.

Figure 4.1 shows all the bit-addressable SFRs and the function of each addressable bit. (Refer to Chapter 2 for more detailed descriptions of the SFR bit functions.)

Bit-Level Boolean Operations

The bit-level Boolean logical opcodes operate on any addressable RAM or SFR bit. The carry flag (C) in the PSW special-function register is the destination for most of the opcodes because the flag can be tested and the program flow changed using instructions covered in Chapter 6.

The following table lists the Boolean bit-level operations.

Mnemonic	Operation
ANL C,b	AND C and the addressed bit; put the result in C
ANL C,/b	AND C and the complement of the addressed bit; put the result in C; the addressed bit is not altered
ORL C,b	OR C and the addressed bit; put the result in C
ORL C,/b	OR C and the complement of the addressed bit; put the result in C; the addressed bit is not altered
CPL C	Complement the C flag
CPL b	Complement the addressed bit
CLR C	Clear the C flag to zero
CLR b	Clear the addressed bit to zero
MOV C,b	Copy the addressed bit to the C flag
MOV b,C	Copy the C flag to the addressed bit
SETB C	Set the flag to one
SETB b	Set the addressed bit to one

Note that no flags, other than the C flag, are affected, unless the flag is an addressed bit.

As is the case for byte-logical operations when addressing ports as destinations, a port bit used as a destination for a logical operation is part of the SFR latch, not the pin. A port bit used as a source *only* is a pin, not the latch. The bit instructions that can use a SFR latch bit are: CLR, CPL, MOV, and SETB.

FIGURES 4.1 Bit-Addressable Control Registers

7	6	5	4	3	2	1	0
CY	AC	F0	RS1	RS0	OV	Reserved	P

PROGRAM STATUS WORD (PSW) SPECIAL FUNCTION REGISTER. BIT ADDRESSES D0h to D7h.

Bit	Function
7	Carry flag
6	Auxiliary carry flag
5	User flag 0
4	Register bank select bit 1
3	Register bank select bit 0
2	Overflow flag
1	Not used (reserved for future)
0	Parity flag

7	6	5	4	3	2	1	0
EA	Reserved	Reserved	ES	ET1	EX1	ET0	EX0

INTERRUPT ENABLE (IE) SPECIAL FUNCTION REGISTER. BIT ADDRESSES A8h TO AFh.

Bit	Function
7	Disables all interrupts
6	Not used (reserved for future)
5	Not used (reserved for future)
4	Serial port interrupt enable
3	Timer 1 overflow interrupt enable
2	External interrupt 1 enable
1	Timer 0 interrupt enable
0	External interrupt 0 enable

EA disables all interrupts when cleared to 0; if EA = 1 then each individual interrupt will be enabled if 1, and disabled if 0.

7	6	5	4	3	2	1	0
*	*	Reserved	PS	PT1	PX1	PT0	PX0

INTERRUPT PRIORITY (IP) SPECIAL FUNCTION REGISTER. BIT ADDRESSES B8h to BFh.

Bit	Function
7	Not implemented
6	Not implemented

Continued

Bit	Function
5	Not used (reserved for future)
4	Serial port interrupt priority
3	Timer 1 interrupt priority
2	External interrupt 1 priority
1	Timer 0 interrupt priority
0	External interrupt 0 priority

The priority bit may be set to 1 (highest) or 0 (lowest).

7	6	5	4	3	2	1	0
TF1	TR1	TF0	TR0	IE1	IT1	IE0	IT0

TIMER/COUNTER CONTROL (TCON) SPECIAL FUNCTION REGISTER. BIT ADDRESSES 88h to 8Fh.

Bit	Function
7	Timer 1 overflow flag
6	Timer run control
5	Timer 0 overflow flag
4	Timer 0 run control
3	External interrupt 1 edge flag
2	External interrupt 1 mode control
1	External interrupt 0 edge flag
0	External interrupt 0 mode control

All flags can be set by the indicated hardware action; the flags are cleared when interrupt is serviced by the processor.

7	6	5	4	3	2	1	0
SM0	SM1	SM2	REN	TB8	RB8	TI	RI

SERIAL PORT CONTROL (SCON) SPECIAL FUNCTION REGISTER. BIT ADDRESSES 98h to 9Fh.

Bit	Function
7	Serial port mode bit 0
6	Serial port mode bit 1
5	Multiprocessor communications enable
4	Receive enable
3	Transmitted bit in modes 2 and 3
2	Received bit in modes 2 and 3
1	Transmit interrupt flag
0	Receive interrupt flag

Bit-level logical operation examples are shown in the following table:

Mnemonic	Operation
SETB 00h	Bit 0 of RAM byte 20h = 1
MOV C,00h	C = 1
MOV 7Fh,C	Bit 7 of RAM byte 2Fh = 1
ANL C,/00h	C = 0; bit 0 of RAM byte 20h = 1
ORL C,00h	C = 1
CPL 7Fh	Bit 7 of RAM byte 2Fh = 0
CLR C	C = 0
ORL C,/7Fh	C = 1; bit 7 of RAM byte 2Fh = 0

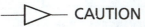

─────── CAUTION ──

Only the SFRs that have been identified as bit addressable may be used in bit operations.

If the destination bit is a port bit, the SFR latch bit is affected, not the pin.

ANL C,/b and ORL C,/b do not alter the addressed bit b.

Rotate and Swap Operations

The ability to rotate data is useful for inspecting bits of a byte without using individual bit opcodes. The A register can be rotated one bit position to the left or right with or without including the C flag in the rotation. If the C flag is not included, then the rotation involves the eight bits of the A register. If the C flag is included, then nine bits are involved in the rotation. Including the C flag enables the programmer to construct rotate operations involving any number of bytes.

The SWAP instruction can be thought of as a rotation of nibbles in the A register. Figure 4.2 diagrams the rotate and swap operations, which are given in the following table:

Mnemonic	Operation
RL A	Rotate the A register one bit position to the left; bit A0 to bit A1, A1 to A2, A2 to A3, A3 to A4, A4 to A5, A5 to A6, A6 to A7, and A7 to A0
RLC A	Rotate the A register and the carry flag, as a ninth bit, one bit position to the left; bit A0 to bit A1, A1 to A2, A2 to A3, A3 to A4, A4 to A5, A5 to A6, A6 to A7, A7 to the carry flag, and the carry flag to A0
RR A	Rotate the A register one bit position to the right; bit A0 to bit A7, A6 to A5, A5 to A4, A4 to A3, A3 to A2, A2 to A1, and A1 to A0
RRC A	Rotate the A register and the carry flag, as a ninth bit, one bit position to the right; bit A0 to the carry flag, carry flag to A7, A7 to A6, A6 to A5, A5 to A4, A4 to A3, A3 to A2, A2 to A1, and A1 to A0
SWAP A	Interchange the nibbles of register A; put the high nibble in the low nibble position and the low nibble in the high nibble position

Note that no flags, other than the carry flag in RRC and RLC, are affected. If the carry is used as part of a rotate instruction, the state of the carry flag should be known before the rotate is done.

FIGURE 4.2   Register A Rotate Operations

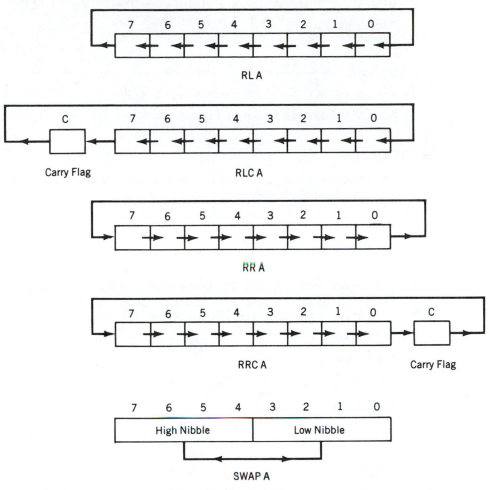

The following table shows examples of rotate and swap operations:

Mnemonic	Operation
MOV A,#0A5h	A = 10100101b = A5h
RR A	A = 11010010b = D2h
RR A	A = 01101001b = 69h
RR A	A = 10110100b = B4h
RR A	A = 01011010b = 5Ah
SWAP A	A = 10100101b = A5h
CLR C	C = 0; A = 10100101b = A5h
RRC A	C = 1; A = 01010010b = 52h
RRC A	C = 0; A = 10101001b = A9h
RL A	A = 01010011b = 53h
RL A	A = 10100110b = A6h

SWAP A	C = 0; A = 01101010b = 6Ah
RLC A	C = 0; A = 11010100b = D4h
RLC A	C = 1; A = 10101000b = A8h
SWAP A	C = 1; A = 10001010b = 8Ah

 CAUTION ───

Know the state of the carry flag when using RRC or RRL.

Rotation and swap operations are limited to the A register.

Example Programs

The programs in this section are written using only opcodes covered to this point in the text. The challenge is to minimize the number of lines of code.

EXAMPLE PROBLEM 4.1

Double the number in register R2, and put the result in registers R3 (high byte) and R4 (low byte).

■ **Thoughts on the Problem** The largest number in R2 is FFh; the largest result is 1FEh. There are at least three ways to solve this problem: Use the MUL instruction (multiply, covered in Chapter 5), add R2 to itself, or shift R2 left one time. The solution that shifts R2 left is as follows:

Mnemonic	**Operation**
MOV R3,#00h	Clear R3 to receive high byte
CLR C	Clear the carry to receive high bit of 2 × R2
MOV A,R2	Get R2 to A
RLC A	Rotate left, which doubles the number in A
MOV R4,A	Put low byte of result in R4
CLR A	Clear A to receive carry
RLC A	The carry bit is now bit 0 of A
MOV R3,A	Transfer any carry bit to R3

 COMMENT ───

Note how the carry flag has to be cleared to a known state before being used in a rotate operation.

EXAMPLE PROBLEM 4.2

OR the contents of ports 1 and 2; put the result in external RAM location 0100h.

■ **Thoughts on the Problem** The ports should be input ports for this problem to make any physical sense; otherwise, we would not know whether to use the pin data or the port SFR latch data.

The solution is as follows:

Mnemonic	**Operation**
MOV A,90h	Copy the pin data from port 1 to A
ORL A,0A0h	OR the contents of A with port 2; results in A
MOV DPTR,#0100h	Set the DPTR to point to external RAM address
MOVX @DPTR,A	Store the result

 **COMMENT**

Any time the port is the source of data, the pin levels are read; when the port is the destination, the latch is written. If the port is both source and destination (read–modify–write instructions), then the latch is used.

EXAMPLE PROBLEM 4.3

Find a number that, when XORed to the A register, results in the number 3Fh in A.

■ **Thoughts on the Problem** Any number can be in A, so we will work backwards:

$$3Fh = A \text{ XOR } N \qquad A \text{ XOR } 3Fh = A \text{ XOR } A \text{ XOR } N = N$$

The solution is as follows:

Mnemonic	Operation
MOV R0,A	Save A in R0
XOR A,#3Fh	XOR A and 3Fh; forming N
XOR A,R0	XOR A and N yielding 3Fh

 **COMMENT**

Does this program work? Let's try several A's and see.

$$A = FFh \qquad A \text{ XOR } 3Fh = C0h \qquad C0h \text{ XOR } FFh = 3Fh$$
$$A = 00h \qquad A \text{ XOR } 3Fh = 3Fh \qquad 3Fh \text{ XOR } 00h = 3Fh$$
$$A = 5Ah \qquad A \text{ XOR } 3Fh = 65h \qquad 65h \text{ XOR } 5Ah = 3Fh$$

Summary

Boolean logic, rotate, and swap instructions are covered in this chapter. Byte-level operations involve each individual bit of a source byte operating on the same bit position in the destination byte; the results are put in the destination, while the source is not changed:

ANL destination,source

ORL destination,source

XRL destination,source

CLR A

CPL A

RR A

RL A

RRC A

RLC A

SWAP A

Bit-level operations involve individual bits found in one area of internal RAM and certain SFRs that may be addressed both by the assigned direct-byte address and eight individual bit addresses. The following Boolean logical operations may be done on each of these addressable bits:

ANL bit

ORL bit

CLR bit

CPL bit

SETB bit

MOV destination bit, source bit

Problems

Write programs that perform the tasks listed using only opcodes that have been discussed in this and previous chapters. Write comments for each line of code and try to use as few lines as possible.

1. Set Port 0, bits 1,3,5, and 7, to one; set the rest to zero.

2. Clear bit 3 of RAM location 22h without affecting any other bit.

3. Invert the data on the port 0 pins and write the data to port 1.

4. Swap the nibbles of R0 and R1 so that the low nibble of R0 swaps with the high nibble of R1 and the high nibble of R0 swaps with the low nibble of R1.

5. Complement the lower nibble of RAM location 2Ah.

6. Make the low nibble of R5 the complement of the high nibble of R6.

7. Make the high nibble of R5 the complement of the low nibble of R6.

8. Move bit 6 of R0 to bit 3 of port 3.

9. Move bit 4 of RAM location 30h to bit 2 of A.

10. XOR a number with whatever is in A so that the result is FFh.

11. Store the most significant nibble of A in both nibbles of register R5; for example, if A = B6h, then R5 = BBh.

12. Store the least significant nibble of A in both nibbles of RAM address 3Ch; for example, if A = 36h, then 3Ch = 66h.

13. Set the carry flag to one if the number in A is even; set the carry flag to zero if the number in A is odd.

14. Treat registers R0 and R1 as 16-bit registers, and rotate them one place to the left; bit 7 of R0 becomes bit 0 of R1, bit 7 of R1 becomes bit 0 of R0, and so on.

15. Repeat Problem 14 but rotate the registers one place to the right.

16. Rotate the DPTR one place to the left; bit 15 becomes bit 0.

17. Repeat problem 16 but rotate the DPTR one place to the right.

18. Shift register B one place to the left; bit 0 becomes a zero, bit 6 becomes bit 7, and so on. Bit 7 is lost.

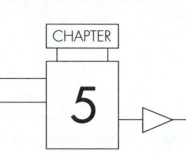

CHAPTER

5

Arithmetic Operations

Chapter Outline

Introduction
Flags
Incrementing and Decrementing
Addition
Subtraction

Multiplication and Division
Decimal Arithmetic
Example Programs
Summary

Introduction

Applications of microcontrollers often involve performing mathematical calculations on data in order to alter program flow and modify program actions. A microcontroller is not designed to be a "number cruncher," as is a general-purpose computer. The domain of the microcontroller is that of controlling events as they change (real-time control). A sufficient number of mathematical opcodes must be provided, however, so that calculations associated with the control of simple processes can be done, in real time, as the controlled system operates. When faced with a control problem, the programmer must know whether the 8051 has sufficient capability to expeditiously handle the required data manipulation. If it does not, a higher performance model must be chosen.

The 24 arithmetic opcodes are grouped into the following types:

Mnemonic	Operation
INC destination	Increment destination by 1
DEC destination	Decrement destination by 1
ADD/ADDC destination,source	Add source to destination without/with carry (C) flag
SUBB destination,source	Subtract, with carry, source from destination
MUL AB	Multiply the contents of registers A and B

DIV AB   Divide the contents of register A by the contents of
                                        register B

DA A   Decimal Adjust the A register

The addressing modes for the destination and source are the same as those discussed in Chapter 3: immediate, register, direct, and indirect.

Flags

A key part of performing arithmetic operations is the ability to store certain results of those operations that affect the way in which the program operates. For example, adding together two one-byte numbers results in a one-byte partial sum, because the 8051 is and eight-bit machine. But it is possible to get a 9-bit result when adding two 8-bit numbers. The ninth bit must be stored also, so the need for a one-bit register, or carry flag in this case, is identified. The program will then have to deal with the ninth bit, perhaps by adding it to a higher order byte in a multiple-byte addition scheme. Similar actions may have to be taken when a larger byte is subtracted from a smaller one. In this case, a borrow is necessary and must be dealt with by the program.

The 8051 has several dedicated latches, or flags, that store results of arithmetic operations. Opcodes covered in Chapter 6 are available to alter program flow based upon the state of the flags. Not all instructions change the flags, but many a programming error has been made by a forgetful programmer who overlooked an instruction that does change a flag.

The 8051 has four arithmetic flags: the carry (C), auxiliary carry (AC), overflow (OV), and parity (P).

Instructions Affecting Flags

The C, AC, and OV flags are arithmetic flags. They are set to 1 or cleared to 0 automatically, depending upon the outcomes of the following instructions. The following instruction set includes *all* instructions that modify the flags and is not confined to arithmetic instructions:

INSTRUCTION MNEMONIC	FLAGS AFFECTED		
ADD	C	AC	OV
ADDC	C	AC	OV
ANL C,direct	C		
CJNE	C		
CLR C	C = 0		
CPL C	C = $\overline{C}$		
DA A	C		
DIV	C = 0		OV
MOV C,direct	C		
MUL	C = 0		OV
ORL C,direct	C		
RLC	C		
RRC	C		
SETB C	C = 1		
SUBB	C	AC	OV

One should remember, however, that the flags are all stored in the PSW. *Any* instruction that can modify a bit or a byte in that register (MOV, SETB, XCH, etc.) changes the flags. This type of change takes conscious effort on the part of the programmer.

A flag may be used for more than one type of result. For example, the C flag indicates a carry out of the lower byte position during addition and indicates a borrow during subtraction. The instruction that *last* affects a flag determines the use of that flag.

The parity flag is affected by every instruction executed. The P flag will be set to a 1 if the number of 1's in the A register is odd and will be set to 0 if the number of 1's is even. All 0's in A yield a 1's count of 0, which is considered to be *even*. Parity check is an elementary error-checking method and is particularly valuable when checking data received via the serial port.

Incrementing and Decrementing

The simplest arithmetic operations involve adding or subtracting a binary 1 and a number. These simple operations become very powerful when coupled with the ability to repeat the operation—that is, to "INCrement" or "DECrement"—until a desired result is reached.[1] Register, Direct, and Indirect addresses may be INCremented or DECremented. No math flags (C, AC, OV) are affected.

The following table lists the increment and decrement mnemonics.

Mnemonic	Operation
INC A	Add a one to the A register
INC Rr	Add a one to register Rr
INC add	Add a one to contents of the direct address
INC @ Rp	Add a one to the contents of the address in Rp
INC DPTR	Add a one to the 16-bit DPTR
DEC A	Subtract a one from register A
DEC Rr	Subtract a one from register Rr
DEC add	Subtract a one from the contents of the direct address
DEC @ Rp	Subtract a one from the contents of the address in register Rp

Note that increment and decrement instructions that operate on a port direct address alter the *latch* for that port.

The following table shows examples of increment and decrement arithmetic operations:

Mnemonic	Operation
MOV A,#3Ah	A = 3Ah
DEC A	A = 39h
MOV R0,#15h	R0 = 15h
MOV 15h,#12h	Internal RAM address 15h = 12h
INC @R0	Internal RAM address 15h = 13h
DEC 15h	Internal RAM address 15h = 12h
INC R0	R0 = 16h
MOV 16h,A	Internal RAM address 16h = 39h
INC @R0	Internal RAM address 16h = 3Ah
MOV DPTR,#12FFh	DPTR = 12FFh
INC DPTR	DPTR = 1300h
DEC 83h	DPTR = 1200h (SFR 83h is the DPH byte)

---

[1] This subject will be explored in Chapter 6.

 **CAUTION**

Remember: *No* math flags are affected.

All 8-bit address contents overflow from FFh to 00h.

DPTR is 16 bits; DPTR overflows from FFFFh to 0000h.

The 8-bit address contents underflow from 00h to FFh.

There is no DEC DPTR to match the INC DPTR.

Addition

All addition is done with the A register as the destination of the result. All addressing modes may be used for the source: an immediate number, a register, a direct address, and an indirect address. Some instructions include the carry flag as an additional source of a single bit that is included in the operation at the *least* significant bit position.

The following table lists the addition mnemonics.

Mnemonic	Operation
ADD A,#n	Add A and the immediate number n; put the sum in A
ADD A,Rr	Add A and register Rr; put the sum in A
ADD A,add	Add A and the address contents; put the sum in A
ADD A,@Rp	Add A and the contents of the address in Rp; put the sum in A

Note that the C flag is set to 1 if there is a carry out of bit position 7; it is cleared to 0 otherwise. The AC flag is set to 1 if there is a carry out of bit position 3; it is cleared otherwise. The OV flag is set to 1 if there is a carry out of bit position 7, but not bit position 6 or if there is a carry out of bit position 6 but not bit position 7, which may be expressed as the logical operation

$$OV = C7 \text{ XOR } C6$$

Unsigned and Signed Addition

The programmer may decide that the numbers used in the program are to be unsigned numbers—that is, numbers that are 8-bit positive binary numbers ranging from 00h to FFh. Alternatively, the programmer may need to use both positive and negative signed numbers.

Signed numbers use bit 7 as a *sign* bit in the most significant byte (MSB) of the *group* of bytes *chosen* by the programmer to represent the largest number to be needed by the program. Bits 0 to 6 of the MSB, and any other bytes, express the magnitude of the number. Signed numbers use a 1 in bit position 7 of the MSB as a negative sign and a 0 as a positive sign. Further, all negative numbers are not in true form, but are in 2's complement form. When doing signed arithmetic, the programmer must *know* how large the largest number is to be—that is, how many bytes are needed for each number.

In signed form, a single byte number may range in size from 10000000b, which is −128d to 01111111b, which is +127d. The number 00000000b is 000d and has a positive sign, so there are 128d negative numbers and 128d positive numbers. The C and OV flags have been included in the 8051 to enable the programmer to use either numbering scheme.

Adding or subtracting unsigned numbers may generate a carry flag when the sum exceeds FFh or a borrow flag when the minuend is less than the subtrahend. The OV flag is not used for unsigned addition and subtraction. Adding or subtracting signed numbers can lead to carries and borrows in a similar manner, and to overflow conditions due to the actions of the sign bits.

Unsigned Addition

Unsigned numbers make use of the carry flag to detect when the result of an ADD operation is a number larger than FFh. If the carry is set to one after an ADD, then the carry can be added to a higher order byte so that the sum is not lost. For instance,

$$
\begin{array}{rl}
95d = & 01011111b \\
\underline{189d =} & \underline{10111101b} \\
284d & 1\ 00011100b = 284d
\end{array}
$$

The C flag is set to 1 to account for the carry out from the sum. The program could add the carry flag to another byte that forms the second byte of a larger number.

Signed Addition

Signed numbers may be added two ways: addition of like signed numbers and addition of unlike signed numbers. If unlike signed numbers are added, then it is not possible for the result to be larger than $-128d$ or $+127d$, and the sign of the result will always be correct. For example,

$$
\begin{array}{rl}
-001d = & 11111111b \\
\underline{+027d =} & \underline{00011011b} \\
+026d & 00011010b = +026d
\end{array}
$$

Here, there is a carry from bit 7 so the carry flag is 1. There is also a carry from bit 6, and the OV flag is 0. For this condition, no action need be taken by the program to correct the sum.

If positive numbers are added, there is the possibility that the sum will exceed $+127d$, as demonstrated in the following example:

$$
\begin{array}{rl}
+100d = & 01100100b \\
\underline{+050d =} & \underline{00110010b} \\
+150d & 10010110b = -106d
\end{array}
$$

Ignoring the sign of the result, the magnitude is seen to be $+22d$ which would be correct if we had some way of accounting for the $+128d$, which, unfortunately, is larger than a single byte can hold. There is no carry from bit 7 and the carry flag is 0; there is a carry from bit 6 so the OV flag is 1.

An example of adding two positive numbers that do not exceed the positive limit is:

$$
\begin{array}{rl}
+045d = & 00101101b \\
\underline{+075d =} & \underline{01001011b} \\
+120d & 01111000b = 120d
\end{array}
$$

Note that there are no carries from bits 6 or 7 of the sum; the carry and OV flags are both 0.

The result of adding two negative numbers together for a sum that does not exceed the negative limit is shown in this example:

$$
\begin{array}{rl}
-030d = & 11100010b \\
\underline{-050d =} & \underline{11001110b} \\
-080d & 10110000b = -080d
\end{array}
$$

Here, there is a carry from bit 7 and the carry flag is 1; there is a carry from bit 6 and the OV flag is 0. These are the same flags as the case for adding unlike numbers; no corrections are needed for the sum.

When adding two negative numbers whose sum does exceed $-128d$, we have

$$
\begin{array}{l}
-070d = \ 10111010b \\
\underline{-070d = \ 10111010b} \\
-140d \quad 01110100b = \ +116d
\end{array}
$$

Or, the magnitude can be interpreted as $-12d$, which is the remainder after a carry out of $-128d$. In this example, there is a carry from bit position 7, and no carry from bit position 6, so the carry and the OV flags are set to 1. The magnitude of the sum is correct; the sign bit must be changed to a 1.

From these examples the programming actions needed for the C and OV flags are as follows:

FLAGS		ACTION
C	**OV**	
0	0	None
0	1	Complement the sign
1	0	None
1	1	Complement the sign

A general rule is that *if the OV flag is set, then complement the sign*. The OV flag also signals that the sum exceeds the largest positive or negative numbers thought to be needed in the program.

Multiple-Byte Signed Arithmetic

The nature of multiple-byte arithmetic for signed and unsigned numbers is distinctly different from single byte arithmetic. Using more than one byte in unsigned arithmetic means that carries or borrows are propagated from low-order to high-order bytes by the simple technique of adding the carry to the next highest byte for addition and subtracting the borrow from the next highest byte for subtraction.

Signed numbers appear to behave like unsigned numbers until the last byte is reached. For a signed number, the seventh bit of the highest byte is the sign; if the sign is negative, then the *entire* number is in 2's complement form.

For example, using a two-byte signed number, we have the following examples:

$$
\begin{array}{l}
+32767d = \ 01111111 \ 11111111b = \ 7FFFh \\
+00000d = \ 00000000 \ 00000000b = \ 0000h \\
-00001d = \ 11111111 \ 11111111b = \ FFFFh \\
-32768d = \ 10000000 \ 00000000b = \ 8000h
\end{array}
$$

Note that the lowest byte of the numbers 00000d and $-32768d$ are exactly alike, as are the lowest bytes for $+32767d$ and $-00001d$.

For multi-byte signed number arithmetic, then, the lower bytes are treated as unsigned numbers. All checks for overflow are done only for the highest order byte that contains the sign. An overflow at the highest order byte is not usually recoverable. The programmer has made a *mistake* and probably has made no provisions for a number larger than planned. Some error acknowledgment procedure, or user notification, should be included in the program if this type of mistake is a possibility.

The preceding examples show the need to add the carry flag to higher order bytes in signed and unsigned addition operations. Opcodes that accomplish this task are similar to the ADD mnemonics: A C is appended to show that the carry bit is added to the sum in bit position 0.

The following table lists the add with carry mnemonics:

Mnemonic	Operation
ADDC A,#n	Add the contents of A, the immediate number n, and the C flag; put the sum in A
ADDC A,add	Add the contents of A, the direct address contents, and the C flag; put the sum in A
ADDC A,Rr	Add the contents of A, register Rr, and the C flag; put the sum in A
ADDC A,@Rp	Add the contents of A, the contents of the indirect address in Rp, and the C flag; put the sum in A

Note that the C, AC, and OV flags behave exactly as they do for the ADD commands.

The following table shows examples of ADD and ADDC multiple-byte signed arithmetic operations:

Mnemonic	Operation
MOV A,#1Ch	A = 1Ch
MOV R5,#0A1h	R5 = A1h
ADD A,R5	A = BDh; C = 0, OV = 0
ADD A,R5	A = 5Eh; C = 1, OV = 1
ADDC A,#10h	A = 6Fh; C = 0, OV = 0
ADDC A,#10h	A = 7Fh; C = 0, OV = 0

 CAUTION

ADDC is normally used to add a carry after the LSB addition in a multi-byte process. ADD is normally used for the LSB addition.

Subtraction

Subtraction can be done by taking the 2's complement of the number to be subtracted, the subtrahend, and adding it to another number, the minuend. The 8051, however, has commands to perform direct subtraction of two signed or unsigned numbers. Register A is the destination address for subtraction. All four addressing modes may be used for source addresses. The commands treat the carry flag as a borrow and always subtract the carry flag as part of the operation.

The following table lists the subtract mnemonics.

Mnemonic	Operation
SUBB A,#n	Subtract immediate number n and the C flag from A; put the result in A
SUBB A,add	Subtract the contents of add and the C flag from A; put the result in A
SUBB A,Rr	Subtract Rr and the C flag from A; put the result in A
SUBB A,@Rp	Subtract the contents of the address in Rp and the C flag from A; put the result in A

Note that the C flag is set if a borrow is needed into bit 7 and reset otherwise. The AC flag is set if a borrow is needed into bit 3 and reset otherwise. The OV flag is set if there is a borrow into bit 7 and not bit 6 or if there is a borrow into bit 6 and not bit 7. As in the case for addition, the OV Flag is the XOR of the borrows into bit positions 7 and 6.

Unsigned and Signed Subtraction

Again, depending on what is needed, the programmer may choose to use bytes as signed or unsigned numbers. The carry flag is now thought of as a borrow flag to account for situations when a larger number is subtracted from a smaller number. The OV flag indicates results that must be adjusted whenever two numbers of unlike signs are subtracted and the result exceeds the planned signed magnitudes.

Unsigned Subtraction

Because the C flag is always subtracted from A along with the source byte, it must be set to 0 if the programmer does not want the flag included in the subtraction. If a multi-byte subtraction is done, the C flag is cleared for the first byte and then included in subsequent higher byte operations.

The result will be in true form, with no borrow if the source number is smaller than A, or in 2's complement form, with a borrow if the source is larger than A. These are *not* signed numbers, as all eight bits are used for the magnitude. The range of numbers is from positive 255d (C = 0, A = FFh) to negative 255d (C = 1, A = 01h).

The following example demonstrates subtraction of larger number from a smaller number:

$$
\begin{array}{r}
015d = 00001111b \\
\text{SUBB}\quad 100d = 01100100b \\
\hline
-085d\ \ 1\ 10101011b = 171d
\end{array}
$$

The C flag is set to 1, and the OV flag is set to 0. The 2's complement of the result is 085d. The reverse of the example yields the following result:

$$
\begin{array}{r}
100d = 01100100b \\
015d = 00001111b \\
\hline
085d\quad 01010101b = 085d
\end{array}
$$

The C flag is set to 0, and the OV flag is set to 0. The magnitude of the result is in true form.

Signed Subtraction

As is the case for addition, two combinations of unsigned numbers are possible when subtracting: subtracting numbers of like and unlike signs. When numbers of like sign are subtracted, it is impossible for the result to exceed the positive or negative magnitude limits of +127d or −128d, so the magnitude and sign of the result do not need to be adjusted, as shown in the following example:

$$
\begin{array}{r}
+100d = 01100100b \qquad \text{(Carry flag = 0 before SUBB)} \\
\text{SUBB}\ +126d = 01111110b \\
\hline
-026d\ \ 1\ 11100110b = -026d
\end{array}
$$

There is a borrow into bit positions 7 and 6; the carry flag is set to 1, and the OV flag is cleared.

The following example demonstrates using two negative numbers:

$$-061d = 11000011b \quad \text{(Carry flag} = 0 \text{ before SUBB)}$$
$$\text{SUBB } \underline{-116d = 10001100b}$$
$$+055d \quad 00110111b = +55d$$

There are no borrows into bit positions 6 or 7, so the OV and carry flags are cleared to zero.

An overflow is possible when subtracting numbers of opposite sign because the situation becomes one of adding numbers of like signs, as can be demonstrated in the following example:

$$-099d = 10011101b \quad \text{(Carry flag} = 0 \text{ before SUBB)}$$
$$\text{SUBB } \underline{+100d = 01100100b}$$
$$-199d \quad 00111001b = +057d$$

Here, there is a borrow into bit position 6 but not into bit position 7; the OV flag is set to 1, and the carry flag is cleared to 0. Because the OV flag is set to 1, the result must be adjusted. In this case, the magnitude can be interpreted as the 2's complement of 71d, the remainder after a carry out of 128d from 199d. The magnitude is correct, and the sign needs to be corrected to a 1.

The following example shows a positive overflow:

$$+087d = 01010111b \quad \text{(Carry flag} = 0 \text{ before SUBB)}$$
$$\text{SUBB } \underline{-052d = 11001100b}$$
$$+139d \quad 10001011b = -117d$$

There is a borrow from bit position 7, and no borrow from bit position 6; the OV flag and the carry flag are both set to 1. Again the answer must be adjusted because the OV flag is set to one. The magnitude can be interpreted as a +011d, the remainder from a carry out of 128d. The sign must be changed to a binary 0 and the OV condition dealt with.

The general rule is that *if the OV flag is set to 1, then complement the sign bit.* The OV flag also signals that the result is greater than $-128d$ or $+127d$.

Again, it must be emphasized: When an overflow occurs in a program, an error has been made in the estimation of the largest number needed to successfully operate the program. Theoretically, the program could resize every number used, but this extreme procedure would tend to hinder the performance of the microcontroller.

Note that for all the examples in this section, it is *assumed* that the carry flag = 0 before the SUBB. The carry flag must be 0 before any SUBB operation that depends upon C = 0 is done.

The following table lists examples of SUBB multiple-byte signed arithmetic operations:

Mnemonic	Operation
MOV 0D0h,#00h	Carry flag = 0
MOV A,#3Ah	A = 3Ah
MOV 45h,#13h	Address 45h = 13h
SUBB A,45h	A = 27h; C = 0, OV = 0
SUBB A,45h	A = 14h; C = 0, OV = 0
SUBB A,#80h	A = 94h; C = 1, OV = 1
SUBB A,#22h	A = 71h; C = 0, OV = 0
SUBB A,#0FFh	A = 72h; C = 1, OV = 0

 **CAUTION**

Remember to set the carry flag to zero if it is not to be included as part of the subtraction operation.

Multiplication and Division

The 8051 has the capability to perform 8-bit integer multiplication and division using the A and B registers. Register B is used solely for these operations and has no other use except as a location in the SFR space of RAM that could be used to hold data. The A register holds one byte of data before a multiply or divide operation, and one of the result bytes after a multiply or divide operation.

Multiplication and division treat the numbers in registers A and B as unsigned. The programmer must devise ways to handle signed numbers.

Multiplication

Multiplication operations use registers A and B as both source and destination addresses for the operation. The unsigned number in register A is multiplied by the unsigned number in register B, as indicated in the following table:

Mnemonic	Operation
MUL AB	Multiply A by B; put the low-order byte of the product in A, put the high-order byte in B

The OV flag will be set if $A \times B >$ FFh. Setting the OV flag does *not* mean that an error has occurred. Rather, it signals that the number is larger than eight bits, and the programmer needs to inspect register B for the high-order byte of the multiplication operation. The carry flag is always cleared to 0.

The largest possible product is FE01h when both A and B contain FFh. Register A contains 01h and register B contains FEh after multiplication of FFh by FFh. The OV flag is set to 1 to signal that register B contains the high-order byte of the product; the carry flag is 0.

The following table gives examples of MUL multiple-byte arithmetic operations:

Mnemonic	Operation
MOV A,#7Bh	A = 7Bh
MOV 0F0h,#02h	B = 02h
MUL AB	A = 00h and B = F6h; OV Flag = 0
MOV A,#0FEh	A = FEh
MUL AB	A = 14h and B = F4h; OV Flag = 1

 **CAUTION**

Note there is no comma between A and B in the MUL mnemonic.

Division

Division operations use registers A and B as both source and destination addresses for the operation. The unsigned number in register A is divided by the unsigned number in register B, as indicated in the following table:

Mnemonic	Operation
DIV AB	Divide A by B; put the integer part of quotient in register A and the integer part of the remainder in B

The OV flag is cleared to 0 unless B holds 00h before the DIV. Then the OV flag is set to 1 to show division by 0. The contents of A and B, when division by 0 is attempted, are undefined. The carry flag is always reset.

Division always results in integer quotients and remainders, as shown in the following example:

$$\frac{A = 213d}{B = 017d} = 12 \text{ (quotient) and 9 (remainder)}$$
$$213 \, [(12 \times 17) + 9]$$

When done in hex:

$$\frac{A = 0D5h}{B = 011h} = C \text{ (quotient) and 9 (remainder)}$$

The following table lists examples of DIV multiple-byte arithmetic operations:

Mnemonic	Operation
MOV A,#0FFh	A = FFh (255d)
MOV 0F0h,#2Ch	B = 2C (44d)
DIV AB	A = 05h and B = 23h [255d = (5 × 44) + 35]
DIV AB	A = 00h and B = 05h [05d = (0 × 35) + 5]
DIV AB	A = 00h and B = 00h [00d = (0 × 5) + 0]
DIV AB	A = ?? and B = ??; OV flag is set to one

 CAUTION ─────────────────────────────────────

The original contents of B (the divisor) are lost.

Note there is no comma between A and B in the DIV mnemonic.

Decimal Arithmetic

Most 8051 applications involve adding intelligence to machines where the hexadecimal numbering system works naturally. There are instances, however, when the application involves interacting with humans, who insist on using the decimal number system. In such cases, it may be more convenient for the programmer to use the decimal number system to represent all numbers in the program.

Four bits are required to represent the decimal numbers from 0 to 9 (0000 to 1001) and the numbers are often called *Binary coded decimal* (BCD) numbers. Two of these BCD numbers can then be packed into a single byte of data.

The 8051 does all arithmetic operations in pure binary. When BCD numbers are being used the result will often be a non-BCD number, as shown in the following example:

$$\begin{array}{r} 49\text{BCD} = 01001001\text{b} \\ +38\text{BCD} = 00111000\text{b} \\ \hline 87\text{BCD} \quad 10000001\text{b} = 81\text{BCD} \end{array}$$

Note that to adjust the answer, an 06d needs to be added to the result.

The opcode that adjusts the result of BCD addition is the decimal adjust A for addition (DA A) command, as shown in the following table:

Mnemonic	Operation
DA A	Adjust the sum of two packed BCD numbers found in A register; leave the adjusted number in A.

The C flag is set to 1 if the adjusted number exceeds 99BCD and set to 0 otherwise. The DA A instruction makes use of the AC flag and the binary sums of the individual binary nibbles to adjust the answer to BCD. The AC flag has no other use to the programmer and no instructions—other than a MOV or a direct bit operation to the PSW—affect the AC flag.

It is important to remember that the DA A instruction assumes the added numbers were in BCD *before* the addition was done. Adding hexadecimal numbers and then using DA A will *not* convert the sum to BCD.

The DA A opcode only works when used with ADD or ADDC opcodes and does not give correct adjustments for SUBB, MUL or DIV operations. The programmer might best consider the ADD or ADDC and DA A as a single instruction and use the pair automatically when doing BCD addition in the 8051.

The following table gives examples of BCD multiple-byte arithmetic operations:

Mnemonic	Operation
MOV A,#42h	A = 42BCD
ADD A,#13h	A = 55h; C = 0
DA A	A = 55h; C = 0
ADD A,#17h	A = 6Ch; C = 0
DA A	A = 72BCD; C = 0
ADDC A,#34h	A = A6h; C = 0
DA A	A = 06BCD; C = 1
ADDC A,#11h	A = 18BCD; C = 0
DA A	A = 18BCD; C = 0

—▷— CAUTION ———————————————————————————————

All numbers used must be in BCD form before addition.

Only ADD and ADDC are adjusted to BCD by DA A.

Example Programs

The challenge of the programs presented in this section is writing them using only opcodes that have been covered to this point in the book. Experienced programmers may long for some of the opcodes to be covered in Chapter 6, but as we shall see, programs can be written without them.

EXAMPLE PROBLEM 5.1

Add the unsigned numbers found in internal RAM locations 25h, 26h, and 27h together and put the result in RAM locations 30h (MSB) and 31h (LSB).

■ **Thoughts on the Problem** The largest number possible is FFh + FFh = 01FEh + FFh = 02FDh, so that two bytes will hold the largest possible number. The MSB will be set to 0 and any carry bit added to it for each byte addition.

To solve this problem, use an ADD instruction for each addition and an ADDC to the MSB for each carry which might be generated. The first ADD will adjust any carry flag which exists before the program starts.

The complete program is shown in the following table:

Mnemonic	Operation
MOV 31h,#00h	Clear the MSB of the result to 0
MOV A,25h	Get the first byte to be added from location 25h
ADD A,26h	Add the second byte found in RAM location 26h
MOV R0,A	Save the sum of the first two bytes in R0
MOV A,#00h	Clear A to 00
ADDC A,31h	Add the carry to the MSB; carry = 0 after this operation
MOV 31h,A	Store MSB
MOV A,R0	Get partial sum back
ADD A,27h	Form final LSB sum
MOV 30h,A	Store LSB
MOV A,#00h	Clear A for MSB addition
ADDC A,31h	Form final MSB
MOV 31h,A	Store final MSB

 COMMENT

Notice how awkward it becomes to have to use the A register for all operations. Jump instructions, which will be covered in Chapter 6, require less use of A.

EXAMPLE PROBLEM 5.2

Repeat problem 5.1 using BCD numbers.

■ **Thoughts on the Problem** The numbers in the RAM locations *must* be in BCD before the problem begins. The largest number possible is 99d + 99d = 198d + 99d = 297d, so that up to two carries can be added to the MSB.

The solution to this problem is identical to that for unsigned numbers, except a DA A must be added after each ADD instruction. If more bytes were added so that the MSB could exceed 09d, then a DA A would also be necessary after the ADDC opcodes.

The complete program is shown in the following table:

Mnemonic	Operation
MOV 31h,#00h	Clear the MSB of the result to 0
MOV A,25h	Get the first byte to be added from location 25h
ADD A,26h	Add the second byte found in RAM location 26h
DA A	Adjust the answer to BCD form
MOV R0,A	Save the sum of the first two bytes in R0
MOV A,#00h	Clear A to 00
ADDC A,31h	Add the carry to the MSB; carry = 0 after this operation
MOV 31h,A	Store MSB
MOV A,R0	Get partial sum back
ADD A,27h	Form final LSB sum
DA A	Adjust the final sum to BCD
MOV 30h,A	Store LSB
MOV A,#00h	Clear A for MSB addition
ADDC A,31h	Form final MSB
MOV 31h,A	Store final MSB

 **COMMENT**

When using BCD numbers, DA A can best be thought of as an integral part of the ADD instructions.

☐ EXAMPLE PROBLEM 5.3

Multiply the unsigned number in register R3 by the unsigned number on port 2 and put the result in external RAM locations 10h (MSB) and 11h (LSB).

■ **Thoughts on the Problem** The MUL instruction uses the A and B registers; the problem consists of MOVes to A and B followed by MOVes to the external RAM. The complete program is shown in the following table:

Mnemonic	Operation
MOV A,0A0h	Move the port 2 pin data to A
MOV 0F0h,R3	Move the data in R3 to the B register
MUL AB	Multiply the data; A has the low order result byte
MOV R0,#11h	Set R0 to point to external RAM location 11h
MOV @R0,A	Store the LSB in external RAM
DEC R0	Decrement R0 to point to 10h
MOV A,0F0h	Move B to A
MOV @R0,A	Store the MSB in external RAM

 **COMMENT**

Again we see the bottleneck created by having to use the A register for all external data transfers.

More advanced programs which do signed math operations and multi-byte multiplication and division will have to wait for the development of Jump instructions in Chapter 6.

Summary

The 8051 can perform all four arithmetic operations: addition, subtraction, multiplication, and division. Signed and unsigned numbers may be used in addition and subtraction; an OV flag is provided to signal programmer errors in estimating signed number magnitudes needed and to adjust signed number results. Multiplication and division use unsigned numbers. BCD arithmetic may be done using the DA A and ADD or ADDC instructions.

The following table lists the arithmetic mnemonics:

Mnemonic	Operation
ADD A, source	Add the source byte to A; put the result in A and adjust the C and OV flags
ADDC A, source	Add the source byte and the carry to A; put the result in A and adjust the C and OV flags
DA A	Adjust the binary result of adding two BCD numbers in the A register to BCD and adjust the carry flag
DEC source	Subtract a 1 from the source; roll from 00h to FFh
DIV AB	Divide the byte in A by the byte in B; put the quotient in A and the remainder in B; set the OV flag to 1 if B = 00h before the division

INC source	Add a 1 to the source; roll from FFh or FFFFh to 00h or 0000h
MUL AB	Multiply the bytes in A and B; put the high-order byte of the result in B, the low-order byte in A; set the OV flag to 1 if the result is > FFh
SUBB A, source	Subtract the source byte and the carry from A; put the result in A and adjust the C and OV flags

Problems

Write programs that perform the tasks listed using only opcodes that have been discussed in this and previous chapters. Use comments on each line of code and try to use as few lines as possible. All numbers may be considered to be unsigned numbers.

1. Add the bytes in RAM locations 34h and 35h; put the result in register R5 (LSB) and R6 (MSB).

2. Add the bytes in registers R3 and R4; put the result in RAM location 4Ah (LSB) and 4Bh (MSB).

3. Add the number 84h to RAM locations 17h (MSB) and 18h (LSB).

4. Add the byte in external RAM location 02CDh to internal RAM location 19h; put the result into external RAM location 00C0h (LSB) and 00C1h (MSB).

5–8. Repeat Problems 1–4, assuming the numbers are in BCD format.

9. Subtract the contents of R2 from the number F3h; put the result in external RAM location 028Bh.

10. Subtract the contents of R1 from R0; put the result in R7.

11. Subtract the contents of RAM location 13h from RAM location 2Bh; put the result in RAM location 3Ch.

12. Subtract the contents of TH0 from TH1; put the result in TL0.

13. Increment the contents of RAM location 13h, 14h, and 15h using indirect addressing only.

14. Increment TL1 by 10h.

15. Increment external RAM locations 0100h and 0200h.

16. Add a 1 to every external RAM address from 00h to 06h.

17. Add a 1 to every external RAM address from 0100h to 0106h.

18. Decrement TL0, TH0, TL1, and TH1.

19. Decrement external RAM locations 0123h and 01BDh.

20. Decrement external RAM locations 45h and 46h.

21. Multiply the data in RAM location 22h by the data in RAM location 15h; put the result in RAM locations 19h (low byte), and 1Ah (high byte).

22. Square the contents of R5; put the result in R0 (high byte), and R1 (low byte).

23. Divide the data in RAM location 3Eh by the number 12h; put the quotient in R4 and the remainder in R5.

24. Divide the number in RAM location 15h by the data in RAM location 16h; put the resulting quotient in external RAM location 7Ch.

25. Divide the data in RAM location 13h by the data in RAM location 14h, then restore the original data in 13h by multiplying the answer by the data in 14h.

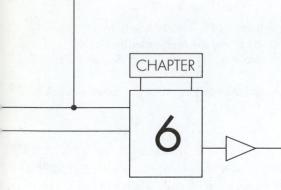

CHAPTER

6

Jump and Call Opcodes

Introduction

The opcodes that have been examined and used in the preceding chapters may be thought of as action codes. Each instruction performs a single operation on bytes of data.

The jumps and calls discussed in this chapter are *decision* codes that alter the flow of the program by examining the results of the action codes and changing the contents of the program counter. A jump permanently changes the contents of the program counter if certain program conditions exist. A call temporarily changes the program counter to allow another part of the program to run. These decision codes make it possible for the programmer to let the program adapt itself, as it runs, to the conditions that exist at the time.

While it is true that computers can't "think" (at least as of this writing), they can make decisions about events that the programmer can foresee, using the following decision opcodes:

Jump on bit conditions

Compare bytes and jump if *not* equal

Decrement byte and jump if zero

Jump unconditionally

Call a subroutine

Return from a subroutine

Jumps and calls may also be generically referred to as "branches," which emphasizes that two divergent paths are made possible by this type of instruction.

The Jump and Call Program Range

A jump or call instruction can replace the contents of the program counter with a new program address number that causes program execution to begin at the code located at the new address. The difference, in bytes, of this new address from the address in the program where the jump or call is located is called the *range* of the jump or call. For example, if a jump instruction is located at program address 0100h, and the jump causes the program counter to become 0120h, then the range of the jump is 20h bytes.

Jump or call instructions may have one of three ranges: a *relative* range of +127d, −128d bytes from the instruction *following* the jump or call instruction; an *absolute* range on the same 2K byte page as the instruction *following* the jump or call; or a *long* range of any address from 0000h to FFFFh, anywhere in program memory. Figure 6.1 shows the relative range of all the jump instructions.

FIGURE 6.1   Jump Instruction Ranges

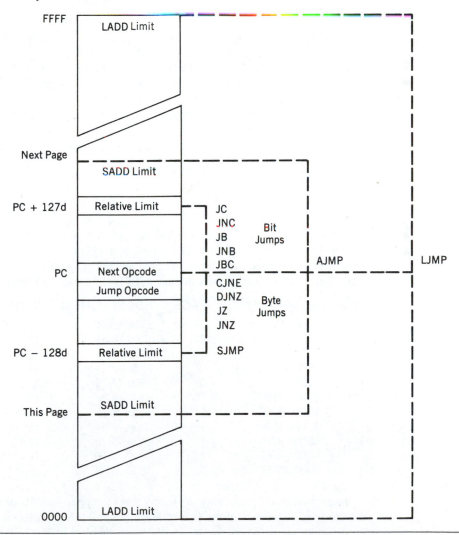

Relative Range

Jumps that replace the program counter contents with a new address that is greater than the address of the instruction *following* the jump by 127d or less than the address of the instruction following the jump by 128d are called *relative* jumps. They are so named because the address that is placed in the program counter is relative to the address where the jump occurs. If the absolute address of the jump instruction changes, then the jump address changes also but remains the same distance away from the jump instruction. The address following the jump is used to calculate the relative jump because of the action of the PC. The PC is incremented to point to the *next* instruction *before* the current instruction is executed. Thus, the PC is set to the following address before the jump instruction is executed, or in the vernacular: "before the jump is taken."

Relative jumping has two advantages. First, only one byte of data need be specified, either in positive format for jumps ahead in the program or in 2's complement negative format for jumps behind. The jump address displacement byte can then be added to the PC to get the absolute address. Specifying only one byte saves program bytes and speeds up program execution. Second, the program that is written using relative jumps can be located anywhere in the program address space without re-assembling the code to generate absolute addresses.

The disadvantage of using relative addressing is the requirement that all addresses jumped be within a range of +127d, −128d bytes of the jump instruction. This range is not a serious problem. Most jumps form program loops over short code ranges that are within the relative address capability. Jumps are the only branch instructions that can use the relative range.

If jumps beyond the relative range are needed, then a relative jump can be done to another relative jump until the desired address is reached. This need is better handled, however, by the jumps that are covered in the next sections.

Short Absolute Range

Absolute range makes use of the concept of dividing memory into logical divisions called "pages." Program memory may be regarded as one continuous stretch of addresses from 0000h to FFFFh. Or, it may be divided into a series of pages of any convenient binary size, such as 256 bytes, 2K bytes, 4K bytes, and so on.

The 8051 program memory is arranged as 2K byte pages, giving a total of 32d (20h) pages. The hexadecimal address of each page is shown in the following table:

PAGE	ADDRESS(HEX)	PAGE	ADDRESS(HEX)	PAGE	ADDRESS(HEX)
00	0000−07FF	0B	5800−5FFF	16	B000−B7FF
01	0800−0FFF	0C	6000−67FF	17	B800−BFFF
02	1000−17FF	0D	6800−6FFF	18	C000−C7FF
03	1800−1FFF	0E	7000−77FF	19	C800−CFFF
04	2000−27FF	0F	7800−7FFF	1A	D000−D7FF
05	2800−2FFF	10	8000−87FF	1B	D800−DFFF
06	3000−37FF	11	8800−8FFF	1C	E000−E7FF
07	3800−3FFF	12	9000−97FF	1D	E800−EFFF
08	4000−47FF	13	9800−9FFF	1E	F000−F7FF
09	4800−4FFF	14	A000−A7FF	1F	F800−FFFF
0A	5000−57FF	15	A800−AFFF		

Inspection of the page numbers shows that the upper five bits of the program counter hold the page *number,* and the lower eleven bits hold the *address* within each page. An absolute address is formed by taking the page number of the instruction *following* the branch and attaching the absolute page range address of eleven bits to it to form the 16-bit address.

Branches on page *boundaries* occur when the jump or call instruction finishes at X7FFh or XFFFh. The next instruction starts at X800h or X000h, which places the jump or call address on the same page as the *next* instruction after the jump or call. The page change presents no problem when branching ahead but could be troublesome if the branch is *backwards* in the program. The assembler should flag such problems as errors, so adjustments can be made by the programmer to use a different type of range.

Absolute range addressing has the same advantages as relative addressing; fewer bytes are needed and the code is relocatable as long as the relocated code begins at the start of a page. Absolute addressing has the advantage of allowing jumps or calls over longer programming distances than does relative addressing.

Long Absolute Range

Addresses that can access the entire program space from 0000h to FFFFh use long range addressing. Long-range addresses require more bytes of code to specify and are relocatable only at the beginning of 64K byte pages. Since we are limited to a nominal ROM address range of 64K bytes, the program must be re-assembled every time a long-range address changes and these branches are not generally relocatable.

Long-range addressing has the advantage of using the entire program address space available to the 8051. It is most likely to be used in large programs.

Jumps

The ability of a program to respond quickly to changes in conditions depends largely upon the number and types of jump instructions available to the programmer. The 8051 has a rich set of jumps that can operate at the bit and byte levels. These jump opcodes are one reason the 8051 is such a powerful microcontroller.

Jumps operate by testing for conditions that are specified in the jump mnemonic. If the condition is *true,* then the jump is taken—that is, the program counter is altered to the address that is part of the jump instruction. If the condition is *false,* then the instruction immediately following the jump instruction is executed because the program counter is not altered. Keep in mind that the condition of *true* does *not* mean a binary 1 and that *false* does *not* mean binary 0. The *condition* specified by the mnemonic is either true or false.

Bit Jumps

Bit jumps all operate according to the status of the carry flag in the PSW or the status of any bit-addressable location. All bit jumps are relative to the program counter.

Jump instructions that test for bit conditions are shown in the following table:

Mnemonic	Operation
JC radd	Jump relative if the carry flag is set to 1
JNC radd	Jump relative if the carry flag is reset to 0
JB b,radd	Jump relative if addressable bit is set to 1
JNB b,radd	Jump relative if addressable bit is reset to 0
JBC b,radd	Jump relative if addressable bit is set, and clear the addressable bit to 0

Note that no flags are affected unless the bit in JBC is a flag bit in the PSW. When the bit used in a JBC instruction is a port bit, the SFR latch for that port is read, tested, and altered.

The following program example makes use of bit jumps:

ADDRESS	MNEMONIC	COMMENT
LOOP:	MOV A,#10h	;A = 10h
	MOV R0,A	;R0 = 10h
ADDA:	ADD A,R0	;add R0 to A
	JNC ADDA	;if the carry flag is 0, then no carry is
		;*true;* jump to address ADDA; jump until A
		;is F0h; the C flag is set to
		;1 on the next ADD and no carry is
		;*false;* do the next instruction
	MOV A,#10h	;A = 10h; do program again using JNB
ADDR:	ADD A,R0	;add R0 to A (R0 already equals 10h)
	JNB 0D7h,ADDR	;D7h is the bit address of the carry flag
	JBC 0D7h,LOOP	;the carry bit is 1; the jump to LOOP
		;is taken, and the carry flag is cleared
		;to 0

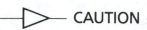 CAUTION

All jump addresses, such as ADDA and ADDR, must be within +127d, −128d of the instruction following the jump opcode.

If the addressable bit is a flag bit and JBC is used, the flag bit will be cleared.

Do not use any label names that are also the names of registers in the 8051. These are called "reserved" words and will cause great agitation in the assembler.

Byte Jumps

Byte jumps—jump instructions that test bytes of data—behave as bit jumps. If the condition that is tested is *true*, the jump is taken; if the condition is *false*, the instruction after the jump is executed. All byte jumps are relative to the program counter.

The following table lists examples of byte jumps:

Mnemonic	Operation
CJNE A,add,radd	Compare the contents of the A register with the contents of the direct address; if they are *not* equal, then jump to the relative address; set the carry flag to 1 if A is less than the contents of the direct address; otherwise, set the carry flag to 0
CJNE A,#n,radd	Compare the contents of the A register with the immediate number n; if they are *not* equal, then jump to the relative address; set the carry flag to 1 if A is less than the number; otherwise, set the carry flag to 0
CJNE Rn,#n,radd	Compare the contents of register Rn with the immediate number n; if they are *not* equal, then jump to the relative address; set the carry flag to 1 if Rn is less than the number; otherwise, set the carry flag to 0
CJNE @Rp,#n,radd	Compare the contents of the address contained in register Rp to the number n; if they are *not* equal, then jump to the relative address; set the carry flag to 1 if the contents of the address in Rp are less than the number; otherwise, set the carry flag to 0

DJNZ Rn,radd	Decrement register Rn by 1 and jump to the relative address if the result is *not* zero; no flags are affected
DJNZ add,radd	Decrement the direct address by 1 and jump to the relative address if the result is *not* 0; no flags are affected unless the direct address is the PSW
JZ radd	Jump to the relative address if A is 0; the flags and the A register are not changed
JNZ radd	Jump to the relative address if A is *not* 0; the flags and the A register are not changed

Note that if the direct address used in a DJNZ is a port, the port SFR is decremented and tested for 0.

Unconditional Jumps

Unconditional jumps do not test any bit or byte to determine whether the jump should be taken. The jump is *always* taken. All jump ranges are found in this group of jumps, and these are the only jumps that can jump to any location in memory.

The following table shows examples of unconditional jumps:

Mnemonic	Operation
JMP @A+DPTR	Jump to the address formed by adding A to the DPTR; this is an unconditional jump and will always be done; the address can be anywhere in program memory; A, the DPTR, and the flags are unchanged
AJMP sadd	Jump to absolute short range address sadd; this is an unconditional jump and is always taken; no flags are affected
LJMP ladd	Jump to absolute long range address ladd; this is an unconditional jump and is always taken; no flags are affected
SJMP radd	Jump to relative address radd; this is an unconditional jump and is always taken; no flags are affected
NOP	Do nothing and go to the next instruction; NOP (no operation) is used to waste time in a software timing loop; or to leave room in a program for later additions; no flags are affected

The following program example uses byte and unconditional jumps:

ADDRESS	MNEMONIC	COMMENT
	.ORG 0100h	;begin program at 0100h
BGN:	MOV A,#30h	;A = 30h
	MOV 50h,#00h	;RAM location 50h = 00h
AGN:	CJNE A,50h,AEQ	;compare A and the contents of 50h in RAM
	SJMP NXT	;SJMP will be executed *if* (50h) = 30h
AEQ:	DJNZ 50h,AGN	;count RAM location 50h down until (50h) =
	NOP	;A; (50h) will reach 30h before 00h
NXT:	MOV R0,#0FFh	;R0 = FFh
DWN:	DJNZ R0,DWN	;count R0 to 00h; loop here until done
	MOV A,R0	;A = R0 = 00h
	JNZ ABIG	;the jump will not be taken
	JZ AZRO	;the jump will be taken

Continued

ADDRESS	MNEMONIC	COMMENT
Continued		
ABIG:	NOP	;this address will not be reached
	.ORG 1000h	;start this segment of program code at ;1000h
AZRO:	MOV A,#08h	;A = 08h (code at 1000,1h)
	MOV DPTR,#1000h	;DPTR = 1000h (code at 1002,3,4h)
	JMP @A+DPTR	;jump to location 1008h (code at 1005h)
	NOP	;(code at 1006h)
	NOP	;(code at 1007h)
HERE:	AJMP AZRO	;(code at 1008h, all code on page 2)

 CAUTION

DJNZ decrements *first, then* checks for 0. A location set to 00h and then decremented goes to FFh, then FEh, and so on, down to 00h.

CJNE does not change the contents of any register or RAM location. It can change the carry flag to 1 if the destination byte is less than the source byte.

There is no zero flag; the JZ and JNZ instructions check the contents of the A register for 0.

JMP @A+DPTR does not change A, DPTR, or any flags.

Calls and Subroutines

The life of a microcontroller would be very tranquil if all programs could run with no thought as to what is going on in the real world outside. However, a microcontroller is specifically intended to interact with the real world and to react, very quickly, to events that require program attention to correct or control.

A program that does not have to deal unexpectedly with the world outside of the microcontroller could be written using jumps to alter program flow as external conditions require. This sort of program can determine external conditions by moving data from the port pins to a location and jumping on the conditions of the port pin data. This technique is called "polling" and requires that the program does not have to respond to external conditions quickly. (Quickly means in microseconds; slowly means in milliseconds.)

Another method of changing program execution is using "interrupt" signals on certain external pins or internal registers to automatically cause a branch to a smaller program that deals with the specific situation. When the event that caused the interruption has been dealt with, the program resumes at the point in the program where the interruption took place. Interrupt action can also be generated using software instructions named *calls*.

Call instructions may be included explicitly in the program as mnemonics or implicitly included using hardware interrupts. In both cases, the call is used to execute a smaller, stand-alone program, which is termed a *routine* or, more often, a *subroutine*.

Subroutines

A *subroutine* is a program that may be used many times in the execution of a larger program. The subroutine could be written into the body of the main program everywhere it is needed, resulting in the fastest possible code execution. Using a subroutine in this manner has several serious drawbacks.

Common practice when writing a large program is to divide the total task among many programmers in order to speed completion. The entire program can be broken into smaller parts and each programmer given a part to write and debug. The main program can then call each of the parts, or subroutines, that have been developed and tested by each individual of the team.

Even if the program is written by one individual, it is more efficient to write an oft-used routine once and then call it many times as needed. Also, when writing a program, the programmer does the main part first. Calls to subroutines, which will be written later, enable the larger task to be defined before the programmer becomes bogged down in the details of the application.

Finally, it is quite common to buy "libraries" of common subroutines that can be called by a main program. Again, buying libraries leads to faster program development.

Calls and the Stack

A call, whether hardware or software initiated, causes a jump to the address where the called subroutine is located. At the end of the subroutine the program resumes operation at the opcode address immediately *following* the call. As calls can be located anywhere in the program address space and used many times, there must be an automatic means of storing the address of the instruction following the call so that program execution can continue after the subroutine has executed.

The *stack* area of internal RAM is used to automatically store the address, called the return address, of the instruction found immediately after the call. The stack pointer register holds the address of the *last* space used on the stack. It stores the return address above this space, adjusting itself upward as the return address is stored. The terms "stack" and "stack pointer" are often used interchangeably to designate the *top* of the stack area in RAM that is pointed to by the stack pointer.

Figure 6.2 diagrams the following sequence of events:

1. A call opcode occurs in the program software, or an interrupt is generated in the hardware circuitry.
2. The return address of the next instruction after the call instruction or interrupt is found in the program counter.
3. The return address bytes are pushed on the stack, *low* byte *first*.
4. The stack pointer is incremented for each push on the stack.
5. The subroutine address is placed in the program counter.
6. The subroutine is executed.
7. A RET (return) opcode is encountered at the end of the subroutine.

FIGURE 6.2   Storing and Retrieving the Return Address

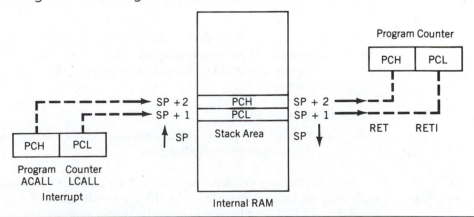

8. Two pop operations restore the return address to the PC from the stack area in internal RAM.

9. The stack pointer is decremented for each address byte pop.

All of these steps are automatically handled by the 8051 hardware. It is the *responsibility* of the programmer to ensure that the subroutine ends in a RET instruction *and* that the stack does not grow up into data areas that are used by the program.

Calls and Returns

Calls use short- or long-range addressing; returns have no addressing mode specified but are always long range. The following table shows examples of call opcodes:

Mnemonic	Operation
ACALL sadd	Call the subroutine located on the same page as the address of the opcode immediately following the ACALL instruction; push the address of the instruction immediately after the call on the stack
LCALL ladd	Call the subroutine located anywhere in program memory space; push the address of the instruction immediately following the call on the stack
RET	Pop two bytes from the stack into the program counter

Note that no flags are affected unless the stack pointer has been allowed to erroneously reach the address of the PSW special-function register.

Interrupts and Returns

As mentioned previously, an *interrupt* is a hardware-generated call. Just as a call opcode can be located within a program to automatically access a subroutine, certain pins on the 8051 can cause a call when external electrical signals on them go to a low state. Internal operations of the timers and the serial port can also cause an interrupt call to take place.

The subroutines called by an interrupt are located at fixed hardware addresses discussed in Chapter 2. The following table shows the interrupt subroutine addresses.

INTERRUPT	ADDRESS (HEX) CALLED
IE0	0003
TF0	000B
IE1	0013
TF1	001B
SERIAL	0023

When an interrupt call takes place, hardware interrupt disable flip-flops are set to prevent another interrupt of the same priority level from taking place until an interrupt return instruction has been executed in the interrupt subroutine. The action of the interrupt routine is shown below.

Mnemonic	Operation
RETI	Pop two bytes from the stack into the program counter and reset the interrupt enable flip-flops

Note that the only difference between the RET and RETI instructions is the enabling of the interrupt logic when RETI is used. RET is used at the ends of subroutines called by an opcode. RETI is used by subroutines called by an interrupt.

The following program example use a call to a subroutine.

ADDRESS	MNEMONIC	COMMENT
MAIN:	MOV 81h,#30h	;set the stack pointer to 30h in RAM
	LCALL SUB	;push address of NOP; PC = #SUB; SP = 32h
	NOP	;return from SUB to this opcode
	. . .	
	. . .	
SUB:	MOV A,#45h	;SUB loads A with 45h and returns
	RET	;pop return address to PC; SP = 30h

 CAUTION

Set the stack pointer above any area of RAM used for additional register banks or data memory.

The stack may only be 128 bytes *maximum;* which limits the number of successive calls with no returns to 64.

Using RETI at the end of a *software* called subroutine may enable the interrupt logic erroneously.

To jump out of a subroutine (not recommended), adjust the stack for the two return address bytes by POPing it twice or by moving data to the stack pointer to reset it to its original value.

Use the LCALL instruction if your subroutines are normally placed at the end of your program.

In the following example of an interrupt call to a routine, timer 0 is used in mode 0 to overflow and set the timer 0 interrupt flag. When the interrupt is generated, the program vectors to the interrupt routine, resets the timer 0 interrupt flag, stops the timer, and returns.

ADDRESS	MNEMONIC	COMMENT
	.ORG 0000h	;begin program at 0000
	AJMP OVER	;jump over interrupt subroutine
	.ORG 000Bh	;put timer 0 interrupt subroutine here
	CLR 8Ch	;stop timer 0; set TR0 = 0
	RETI	;return and enable interrupt structure
	.	
	.	
OVER:	MOV 0A8h,#82h	;enable the timer 0 interrupt in the IE
	MOV 89h,#00h	;set timer operation, mode 0
	MOV 8Ah,#00h	;clear TL0
	MOV 8Ch,#00h	;clear TH0
	SET 8Ch	;start timer 0; set TR0 = 1

```
;
;
;
;the program will continue on and be interrupted when the timer has
;timed out
```

 CAUTION

The programmer must enable any interrupt by setting the appropriate enabling bits in the IE register.

Example Problems

We now have all of the tools needed to write powerful, compact programs. The addition of the decision jump and call opcodes permits the program to alter its operation as it runs.

EXAMPLE PROBLEM 6.1

Place any number in internal RAM location 3Ch and increment it until the number equals 2Ah.

■ **Thoughts on the Problem** The number can be incremented and then tested to see whether it equals 2Ah. If it does, then the program is over; if not, then loop back and increment the number again.

Three methods can be used to accomplish this task.

■ **Method 1:**

ADDRESS	MNEMONIC	COMMENT
ONE:	CLR C	;this program will use SUBB to detect equality
	MOV A,#2Ah	;put the target number in A
	SUBB A,3Ch	;subtract the contents of 3Ch; C is cleared
	JZ DONE	;if A = 00h, then the contents of 3Ch = 2Ah
	INC 3Ch	;if A is not zero, then loop until it is
	SJMP ONE	;loop to try again
DONE:	NOP	;when finished, jump here and continue

━━▷━━ COMMENT ━━━━━━━━━━━━━━━━━━━━━━━━━━━━━━

As there is no compare instruction for the 8051, the SUBB instruction is used to compare A against a number. The SUBB instruction subtracts the C flag also, so the C flag has to be cleared before the SUBB instruction is used.

■ **Method 2:**

ADDRESS	MNEMONIC	COMMENT
TWO:	INC 3Ch	;incrementing 3Ch first saves a jump later
	MOV A,#2Ah	;this program will use XOR to detect equality
	XRL A,3Ch	;XOR with the contents of 3Ch; if equal, A = 00h
	JNZ TWO	;this jump is the reverse of program one
	NOP	;finished when the jump is *false*

━━▷━━ COMMENT ━━━━━━━━━━━━━━━━━━━━━━━━━━━━━━

Many times if the loop is begun with the action that is to be repeated until the loop is satisfied, only one jump, which repeats the loop, is needed.

■ **Method 3:**

ADDRESS	MNEMONIC	COMMENT
THREE:	INC 3Ch	;begin by incrementing the direct address
	MOV A,#2Ah	;this program uses the very efficient CJNE
	CJNE A,3Ch,THREE	;jump if A and (3Ch) are not equal
	NOP	;all done

━━▷━━ COMMENT ━━━━━━━━━━━━━━━━━━━━━━━━━━━━━━

CJNE combines a compare and a jump into one compact *instruction*.

EXAMPLE PROBLEM 6.2

The number A6h is placed somewhere in external RAM between locations 0100h and 0200h. Find the address of that location and put that address in R6 (LSB) and R7 (MSB).

■ **Thoughts on the Problem** The DPTR is used to point to the bytes in external memory, and CJNE is used to compare and jump until a match is found.

ADDRESS	MNEMONIC	COMMENT
	MOV 20h,#0A6h	;load 20h with the number to be found
	MOV DPTR, #00FFh	;start the DPTR below the first address
MOR:	INC DPTR	;increment first and save a jump
	MOVX A,@DPTR	;get a number from external memory to A
	CJNE A,20h,MOR	;compare the number against (20h) and
		;loop to MOR if not equal
	MOV R7,83h	;move DPH byte to R7
	MOV R6,82h	;move DPL byte to R6; finished

 COMMENT

This program might loop forever unless we know the number will be found; a check to see whether the DPTR has exceeded 0200h can be included to leave the loop if the number is not found before DPTR = 0201h.

EXAMPLE PROBLEM 6.3

Find the address of the first two internal RAM locations between 20h and 60h which contain consecutive numbers. If so, set the carry flag to 1, else clear the flag.

■ **Thoughts on the Problem** A check for end of memory will be included as a Called routine, and CJNE and a pointing register will be used to search memory.

ADDRESS	MNEMONIC	COMMENT
	MOV 81h,#65h	;set the stack above memory area
	MOV R0,#20h	;load R0 with address of memory start
NXT:	MOV A,@R0	;get first number
	INC A	;increment and compare to next number
	MOV 1Fh,A	;store incremented number at 1Fh
	INC R0	;point to next number
	CALL DUN	;see if R0 greater than 60h
	JNC THRU	;DUN returns C = 0 if over 60h
	MOV A,@R0	;get next number
	CJNE A,1Fh,NXT	;if not equal then look at next pair
	SETB 0D7h	;set the carry to 1; finished
THRU:	SJMP THRU	;jump here if beyond 60h
DUN:	PUSH A	;save A on the stack
	CLR C	;clear the carry
	MOV A,#61h	;use XOR as a compare
	XRL A,R0	;A will be 0 if equal
	JNZ BCK	;if not 0 then continue
	RET	;A 0, signal calling routine
BCK:	POP A	;get A back
	CPL C	;A not 0, set C to indicate not done
	RET	

 COMMENT

Set the stack pointer to put the stack out of the memory area in use.

Summary

Jumps

Jumps alter program flow by replacing the PC counter contents with the address of the jump address. Jumps have the following ranges:

Relative: up to PC +127 bytes, PC −128 bytes away from the PC

Absolute short: anywhere on a 2K-byte page

Absolute long: anywhere in program memory

Jump opcodes can test an individual bit, or a byte, to check for conditions that make the program jump to a new program address. The bit jumps are shown in the following table:

INSTRUCTION TYPE	RESULT
JC radd	Jump relative if carry flag set to 1
JNC radd	Jump relative if carry flag cleared to 0
JB b,radd	Jump relative if addressable bit set to 1
JNB b,radd	Jump relative if addressable bit cleared to 0
JBC b,radd	Jump relative if addressable bit set to 1 and clear bit to 0

Byte jumps are shown in the following table:

INSTRUCTION TYPE	RESULT
CJNE destination,source,address	Compare destination and source; jump to address if *not* equal
DJNZ destination,address	Decrement destination by one; jump to address if the result is *not* zero
JZ radd	Jump A = 00h to relative address
JNZ radd	Jump A > 00h to relative address

Unconditional jumps make no test and are always made. They are shown in the following table:

INSTRUCTION TYPE	RESULT
JMP @A+DPTR	Jump to 16-bit address formed by adding A to the DPTR
AJMP sadd	Jump to absolute short address
LJMP ladd	Jump to absolute long address
SJMP radd	Jump to relative address
NOP	Do nothing and go to next opcode

Call and Return

Software calls may use short- and long-range addressing; returns are to any long-range address in memory. Interrupts are calls forced by hardware action and call subroutines located at predefined addresses in program memory. The following table shows calls and returns:

INSTRUCTION TYPE	RESULT
ACALL sadd	Call the routine located at absolute short address
LCALL ladd	Call the routine located at absolute long address
RET	Return to anywhere in the program at the address found on the top two bytes of the stack
RETI	Return from a routine called by a hardware interrupt and reset the interrupt logic

Problems

Write programs for each of the following problems using as few lines of code as you can. Place comments on each line of code.

1. Put a random number in R3 and increment it until it equals E1h.

2. Put a random number in address 20h and increment it until it equals a random number put in R5.

3. Put a random number in R3 and decrement it until it equals E1h.

4. Put a random number in address 20h (LSB) and 21h (MSB) and decrement them as if they were a single 16-bit counter until they equal random numbers in R2 (LSB) and R3 (MSB).

5. Random unsigned numbers are placed in registers R0 to R4. Find the largest number and put it in R6.

6. Repeat Problem 5, but find the smallest number.

7. If the lower nibble of any number placed in A is larger than the upper nibble, set the C flag to one; otherwise clear it.

8. Count the number of ones in any number in register B and put the count in R5.

9. Count the number of zeroes in any number in register R3 and put the count in R5.

10. If the signed number placed in R7 is negative, set the carry flag to 1; otherwise clear it.

11. Increment the DPTR from any initialized value to ABCDh.

12. Decrement the DPTR from any initialized value to 0033h, as a 16-bit register.

13. Use R4 (LSB) and R5 (MSB) as a single 16-bit counter, and decrement the pair until they equal 0000h.

14. Get the contents of the PC to the DPTR.

15. Get the contents of the DPTR to the PC.

16. Get any two bytes you wish to the PC.

17. Write a simple subroutine, call it, and jump back to the calling program after adjusting the stack pointer.

18. Put one random number in R2 and another in R5. Increment R2 and decrement R5 until they are equal.

19. Fill external memory locations 100h to 200h with the number AAh.

20. Transfer the data in internal RAM locations 10h to 20h to internal RAM locations 30h to 40h.

21. Set every third byte in internal RAM from address 20h to 7Fh to FFh.

22. Count the number of bytes in external RAM locations 100h to 200h that are greater than the random unsigned number in R3 *and* less than the random unsigned number in R4. Use registers R6 (LSB) and R7 (MSB) to hold the count.

23. Assuming the crystal frequency is 10 megahertz, write a program that will use timer 1 to interrupt the program after a delay of 2 ms.

24. Put the address of every internal RAM byte from 50h to 70h in the address; for instance, internal RAM location 6Dh would contain 6Dh.

25. Put the byte AAh in all internal RAM locations from 20h to 40h, then read them back and set the carry flag to 1 if any byte read back is not AAh.

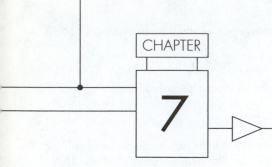

CHAPTER

7

An 8051 Microcontroller Design

Chapter Outline

Introduction
A Microcontroller Specification
A Microcontroller Design
Testing the Design

Timing Subroutines
Lookup Tables for the 8051
Serial Data Transmission
Summary

Introduction

In this chapter a hardware configuration for an 8051 microcontroller, which will be used for all of the example applications in Chapters 8 and 9, is defined. Programs that check the initial prototype of the design (debugging programs) are given in this chapter, followed by several common subroutines that can be used by programs in succeeding chapters.

The design of the microcontroller begins with an identified need and a blank piece of paper or computer screen. The evolution of the microcontroller follows these steps:

1. Define a specification.
2. Design a microcontroller system to this specification.
3. Write programs that will assist in checking the design.
4. Write several common subroutines and test them.

The most important step is the first one. If the application is for high-volume production (greater than 10,000 units), then the task must be very carefully analyzed. A precise or "tight" specification is evolved for what will become a major investment in factory-programmed parts. As the volume goes down for any particular application, the specifications become more general as the designers attempt to write a specification that might fit a wider range of applications.

The list leaves out a few real-world steps, most notably the redesign of the micro-controller after it is discovered that the application has grown beyond the original specification or, as is more common, the application was not well understood in the beginning. Experienced designers learn to add a little "fat" to the specification in anticipation of the inexorable need for "one more bit of I/O and one more kilobyte of memory."

A Microcontroller Specification

A typical outline for a microcontroller design might read as follows:

"A requirement exists for an intelligent controller for real-time control and data monitoring applications. The controller is part of a networked system of identical units that are connected to a host computer through a serial data link. The controller is to be produced in low volumes, typically less than one thousand units for any particular application, and it must be low cost."

The 8051 family is chosen for the following reasons:

Low part cost

Multiple vendors

Available in NMOS and CMOS technologies

Software tools available and inexpensive

High-level language compilers available

The first three items are very important from a production cost standpoint. The software aids available reduce first costs and enable projects to be completed in a timely manner.

The low-volume production requirement and the need for changing the program to fit particular applications establish the necessity of using external EPROM to hold the application program. In turn, ports 0 (AD0–AD7) and 2 (A8–A15) must be used for interfacing to the external ROM and will not be available for I/O.

Because one possible use of the controller will be to gather data, RAM beyond that available internally may be needed. External RAM is added for this eventuality. The immediate consequence of this decision is that port 3 bits 6 ($\overline{\text{WR}}$) and 7 ($\overline{\text{RD}}$) are needed for the external RAM and are not available for I/O. External memory uses the 28-pin standard configuration, which enables memories as large as 64K to be inserted in the memory sockets.

Commercially available EPROM parts that double in size beginning at 2K bytes can be purchased. The minimum EPROM size selected is 8K and the maximum size is 64K. These choices reflect the part sizes that are most readily available from vendors and parts that are now beginning to enter high-volume production.

Static RAM parts are available in 2K, 8K, and 32K byte sizes; again, the RAM sizes are chosen to be 8K or 32K to reflect commercial realities. The various memory sizes can be incorporated by including jumpers for the additional address lines needed by larger memories and pullup resistors to enable alternate pin uses on smaller memories.

The serial data needs can be handled by the internal serial port circuitry. Once again, two more I/O pins of port 3 are used: bits 3.0 (RXD) and 3.1 (TXD). We are left with all of port 1 for general-purpose I/O and port 3 pins 2–5 for general-purpose I/O or for external interrupts and timing inputs.

Note that rapid loss of I/O capability occurs as the alternate port functions are used and should be expected unless volumes are high enough to justify factory-programmed parts.

The handicap is not as great as it appears, however; two methods exist that are commonly used to expand the I/O capability of any computer application: port I/O and memory-mapped I/O.

Finally, we select a 16 megahertz crystal to take advantage of the latest high-speed devices available, and the specification is complete. To summarize, we have

80C31-1 (ROMless) microcontroller

64K bytes of external EPROM

32K bytes of external RAM

8 general-purpose I/O lines

4 general-purpose or programmable I/O lines

1 full-duplex serial port

16 megahertz crystal clock

Now that the specification is complete, the design can be done.

A Microcontroller Design

The final design, shown in Figure 7.1, is based on the external memory circuit found in Chapter 2. Any I/O circuitry needed for a particular application will be added to the basic design as required. A design may be done in several ways; the choices made for this design are constrained by cost and the desire for flexibility.

External Memory and Memory Space Decoding

External memory is added by using port 0 as a data and low-order address bus, and port 2 as a high-order address bus. The data and low addresses are time multiplexed on port 0. An external 373 type address latch is connected to port 0 to store the low address byte whenever external memory is accessed. The low-order address is gated into the transparent latch by the ALE pulse from the 8051. Port 0 then becomes a bidirectional data bus during the read or write phase of a machine cycle.

RAM and ROM are addressed by entirely different control lines from the 8051: $\overline{PSEN}$ for the ROM and $\overline{WR}$ or $\overline{RD}$ for the RAM. The result is that each occupies one of two parallel 64 kilobyte address spaces. The decoding problem becomes one of simply adding suitable jumpers and pullup resistors so that the user can insert the memory capacity needed. Jumpers are inserted so that the correct address line reaches the memory pin or the pin is pulled high as required by the memory used. The jumper table in Figure 7.1 for the EPROM and RAM memories that can be inserted in the memory sockets shows the jumper configuration. Figure 7.2 graphically demonstrates the relative sizes of the internal and external memories available to the programmer.

Reset and Clock Circuits

The 8051 uses an active high reset pin. The reset input must go high for two machine cycles when power is first applied and then sink low. The simple RC circuit used here will supply system voltage (Vcc) to the reset pin until the capacitor begins to charge. At a threshold of about 2.5 V, the reset input reaches a low level, and the system begins to run. Internal reset circuitry has hysteresis necessitated by the slow fall time of the RC circuit.

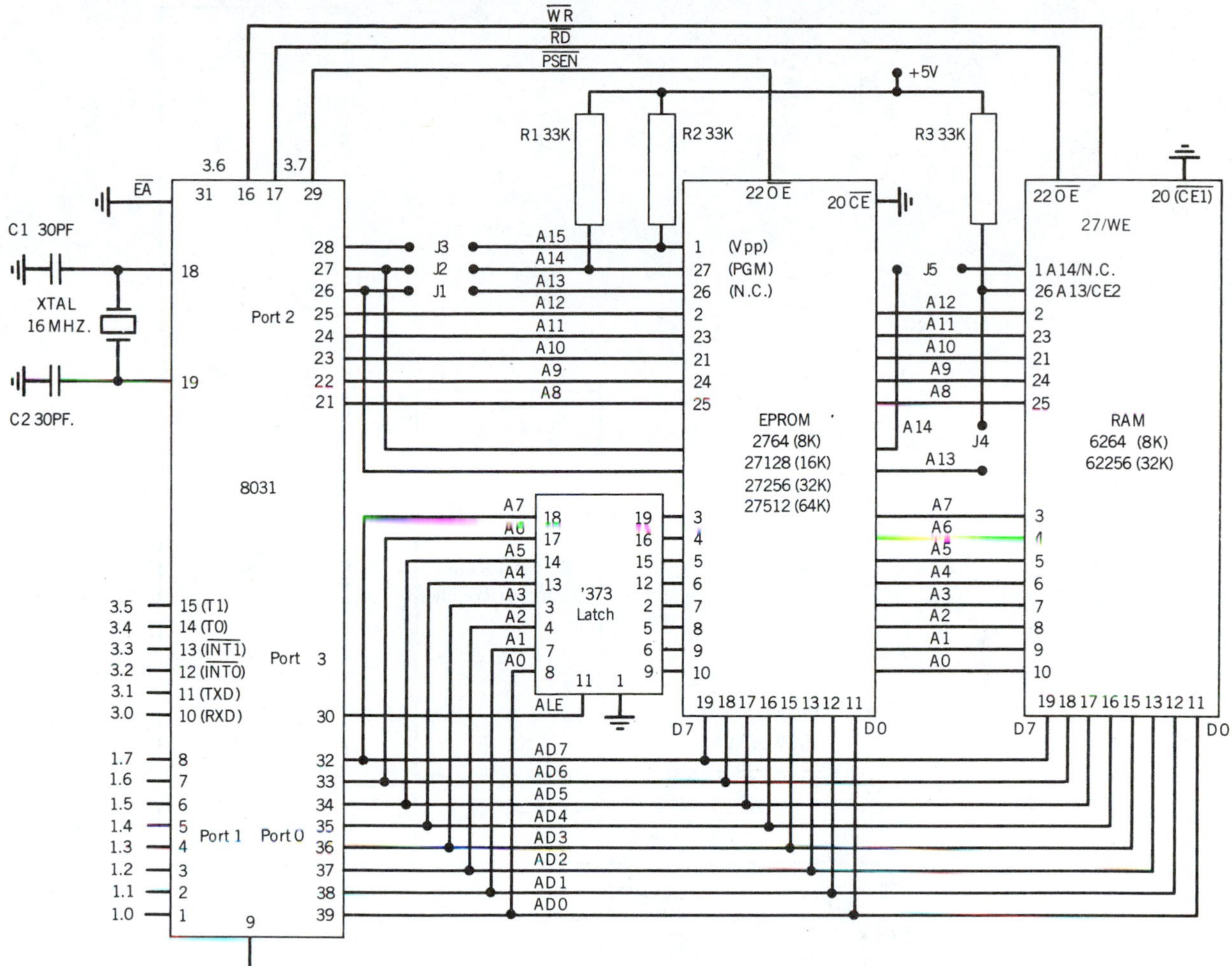

FIGURE 7.1 8031 Microcontroller with External ROM and RAM

Jumper Table

Memory Size	EPROM	RAM
8K	None	None
16K	J1	NA
32K	J1 J2	J4 J5
64K	J1 J2 J3	NA

The addition of a reset button enables the user to reset the system without having to turn power off and on.

The clock circuit of Chapter 2 is added, and the design is finished.

Expanding I/O

Ports 1 and 3 can be used to form small control and bidirectional data buses. The data buses can interface with additional external circuits to expand I/O up to any practical number of lines.

FIGURE 7.2   8031 Memory Sizes

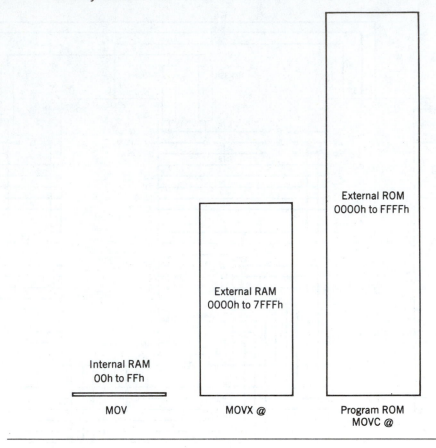

There are many popular families of programmable port chips. The one chosen here is the popular 8255 programmable interface adaptor, which is available from a number of vendors. Details on the full capabilities of the 8255 are given in Appendix D. The 8255 has an internal mode register to which control words are written by the host computer. These control words determine the actions of the 8255 ports, named A, B, and C, enabling them to act as input ports, output ports, or some combination of both.

Figure 7.3 shows a circuit that adds an 8255 port expansion chip to the design. The number of ports is now three 8-bit ports for the system. The penalty paid for expanding I/O in this manner is a reduction in speed that occurs due to the overhead time needed to write control bits to ports 1 and 3 before the resulting I/O lines selected can be accessed. The advantage of using I/O port expansion is that the entire range of 8051 instructions can be used to access the added ports via ports 1 and 3.

Memory-Mapped I/O

The same programmable chip used for port expansion can also be added to the RAM memory space of the design, as shown in Figure 7.4. The present design uses only 32K of

FIGURE 7.3 Expanding I/O Using 8031 Ports

8031		8255	
Port 3	15 — $\overline{CS}$	6	4 — PA0
	14 — Reset		3 — PA1
	13 — A0	35	2 — PA2
	12 — A1	9	1 — PA3
	11 — $\overline{RD}$	8	40 — PA4
	10 — $\overline{WR}$	5	39 — PA5
		36	38 — PA6
			37 — PA7
Port 1	1 — D0	34	18 — PB0
	2 — D1	33	19 — PB1
	3 — D2	32	20 — PB2
	4 — D3	31	21 — PB3
	5 — D4	30	22 — PB4
	6 — D5	29	23 — PB5
	7 — D6	28	24 — PB6
	8 — D7	27	25 — PB7
			14 — PC0
			15 — PC1
			16 — PC2
			17 — PC3
			13 — PC4
			12 — PC5
			11 — PC6
			10 — PC7

the permitted 64K of RAM address space; the upper 32K is vacant. The port chip can be addressed any time A15 is high (8000h or above), and the 32K RAM can be addressed whenever A15 is low (7FFFh and below). This decoding scheme requires only the addition of an inverter to decode the memory space for RAM and I/O.

Should more RAM be added to the design, a comprehensive memory-decoding scheme will require the use of a programmable array-type decoder to reserve some portion of memory space for the I/O port chips. Figure 7.4 shows a design that permits the addition of three memory-mapped port chips at addresses FFF0h–FFF3h, FFF4h–FFF7h, FFF8h–FFFBh, and FFFCh–FFFFh. RAM is addressable from 0000h to FFEFh.

Memory-mapped I/O has the advantage of not using any of the 8051 ports. Disadvantages include the loss of memory space for RAM that is used by the I/O address space, or the addition of memory decoding chips in order to limit the RAM address space loss. Programming overhead is about the same as for port I/O because only the cumbersome MOVX instruction may be used to access the memory-mapped I/O.

For both types of I/O expansion, the cost of the system begins to mount. At some point, a conventional microprocessor, with a rich set of I/O and memory instructions, may become a more economical choice.

FIGURE 7.4 Expanding I/O Using Memory Mapping

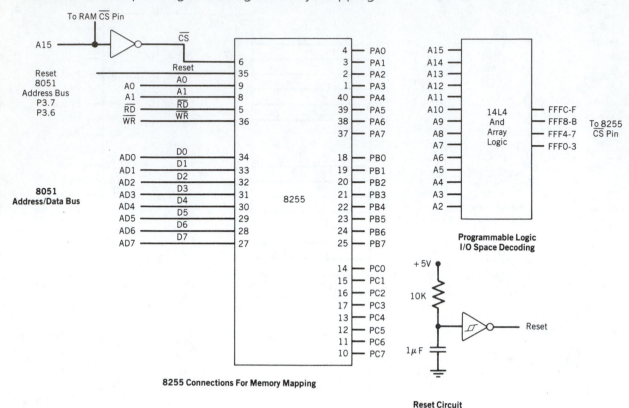

8255 Connections For Memory Mapping

Programmable Logic I/O Space Decoding

Reset Circuit

Part Speed

One consideration, that does not appear on the design drawings, is the selection of parts that will work at the system speeds determined by the crystal frequency. All memory parts are priced according to the nanosecond of access time. The longer the access time (the time it takes for a byte of data to be read or written from or to the device after the address is valid), the cheaper the part. For our design, the required memory circuit timing is not critical as long as we use frequencies below 12 MHz. These times are *totally* determined by the selection of the crystal frequency, and the designer *must* choose memory parts that are fast enough to keep up with the microcontroller at the chosen frequency. For our example, EPROMS with maximum access times of 150 ns and RAM with access times of 400 ns must be used. These access times are representative of standard commercial types currently available at the low end of the cost spectrum. These times are worst-case times; actual access times are at least 30 percent longer.

Other parts, such as the '373 type latch can be any family from LSTTL to HCMOS. The speeds of these parts far exceed the speed of the 8051.

Production Concerns

The design omits many features that would be incorporated by a design-manufacturing team. Chief among these are the inclusion of test-points, LED indicators, and other items that should be added to enhance manufacturing and field service of the microcontroller.

These concerns are well beyond the scope of this book, but the wise designer always ensures that the legitimate concerns of the technical, manufacturing, and service departments are addressed.

Testing the Design

Once the hardware has been assembled, it is necessary to verify that the design is correct and that the prototype is built to the design drawing. This verification of the design is done by running several small programs, beginning with the most basic program and building on the demonstrated success of each.

Crystal Test

The initial test is to ensure that both the crystal and the reset circuit are working. The 8051 is inserted in the circuit, and the ALE pulse is checked with an oscilloscope to verify that the ALE frequency is 1/6 of the crystal frequency. Next, the reset button is pushed, and all ports are checked to see that they are in the high (input) state.

ROM Test

The most fundamental program test is to ensure that the microcontroller can fetch and execute programs from the EPROM. Code byte fetching can be tested by verifying that each address line of the ROM is properly wired by using a series of repeated jump instructions that exercise all of the address lines. The test used here will jump to addresses that are a power of two. Only one address line will be high, and all of the rest will be low. The address pattern tests for proper wiring of each address line and also checks for shorts between any two lines.

If the test is successful, the program stops at the highest possible ROM address. The address bus can then be checked with a logic probe to verify that the highest address has been reached. Correct operation is indicated by the highest order address bus bit, which will appear constant. If not, the probe will blink indicating random program fetches.

The test is run by inserting the '373 latch, the programmed 64K EPROM, inserting jumpers 1–3 and resetting the 8051. The test can be stopped at any address by jumping to that address, as is done in the last statement in the following ROM test program:

ADDRESS	MNEMONIC	COMMENT
	.org 0000h	;start at the bottom of ROM
begin:	ljmp add2	;test address lines A0 and A1
	.org 0004h	;next jump at address 0004h (A2)
add2:	ljmp add3	;test address line A2
	.org 0008h	;next jump at address 0008h (A3)
add3:	ljmp add4	;test address line A3
	.org 0010h	;next jump at address 0010h (A4)
add4:	ljmp add5	;test address line A4
	.org 0020h	;next jump at address 0020h (A5)
add5:	ljmp add6	;test address line A5
	.org 0040h	;next jump at address 0040h (A6)
add6:	ljmp add7	;test address line A6
	.org 0080h	;next jump at address 0080h (A7)

Continued

ADDRESS	MNEMONIC	COMMENT

Continued

```
add7:       ljmp add8       ;test address line A7
            .org 0100h      ;next jump at address 0100h (A8)
add8:       ljmp add9       ;test address line A8
            .org 0200h      ;next jump at address 0200h (A9)
add9:       ljmp add10      ;test address line A9
            .org 0400h      ;next jump at address 0400h (A10)
add10:      ljmp add11      ;test address line A10
            .org 0800h      ;next jump at address 0800h (A11)
add11:      ljmp add12      ;test address line A11
            .org 1000h      ;next jump at address 1000h (A12)
add12:      ljmp add13      ;test address line A12
            .org 2000h      ;next jump at address 2000h (A13)
add13:      ljmp add14      ;test address line A13
            .org 4000h      ;last jump at address 4000h (A14)
add14:      ljmp add15      ;test address line A14
            .org 8000h      ;test address line A15 and remain here
add15:      .ljmp add15     ;jump here in a loop
            .end            ;assembler use
;
;This address, A15, will remain latched while A2—A14 will
;remain low. A0 and A1 will vary as the bytes of the jump
;instruction are fetched.
```

Inspection of the listing for this program in Figure 7.5 shows that all the address lines are exercised.

RAM Test

Once sure of the ability of the microcontroller to execute code, the RAM can be checked. A common test is to write a so-called checkerboard pattern to RAM—that is, an alternating pattern of 1 and 0 in memory. Writing bytes of 55h or AAh will generate such a pattern.

The next program writes this pattern to external RAM, then reads the pattern back and checks each byte read back against the byte that was written. If a check fails, then the address where the failure occurred is in the DPTR register. Port 1 and the free bits of port 3 can then be used to indicate the contents of DPTR.

There are 14 bits available using these ports (the serial port is not in use now, so bits 3.0 and 3.1 are free), and 15 are needed to express a 32K address range. The program will test a range of 8K bytes at a time, using 13 bits to hold the 8K address upon failure. Four versions have to be run to cover the entire RAM address space. If the test is passed, then bit 14 (port 3.5) is a 1. If the test fails, then bit 14 is a 0, and the other 13 bits hold the address (in the 8K page) at which the failure occurred.

Interestingly, this test does not check for correct wiring of the RAM address lines. As long as all address lines end on some valid address, the test will work. A wiring check requires that a ROM be programmed with some unique pattern at each address that is a power of two and read using a check program that inspects each unique address location for a unique pattern.

The RAM test program is listed on the following page.

FIGURE 7.5 Assembled ROM Check Program

```
0000                           .org 0000h      ;start at the bottom of ROM
0000 020004  begin:   ljmp add2       ;test address lines A0 and A1
0004                           .org 0004h      ;next jump at address 0004h (A2)
0004 020008  add2:    ljmp add3       ;test address line A2
0008                           .org 0008h      ;next jump at address 0008h (A3)
0008 020010  add3:    ljmp add4       ;test address line A3
0010                           .org 0010h      ;next jump at address 0010h (A4)
0010 020020  add4:    ljmp add5       ;test addresss line A4
0020                           ,org 0020h      ;next jump at address 0020h (A5)
0020 020040  add5:    ljmp add6       ;test address line A5
0040                           .org 0040h      ;next jump at address 0040h (A6)
0040 020080  add6:    ljmp add7       ;test address line A6
0080                           .org 0080h      ;next jump at address 0080h (A7)
0080 020100  add7:    ljmp add8       ;test address line A7
0100                           .org 0100h      ;next jump at address 0100h (A8)
0100 020200  add8:    ljmp add9       ;test address line A8
0200                           .org 0200h      ;next jump at address 0200h (A9)
0200 020400  add9:    ljmp add10      ;test address line A9
0400                           .org 0400h      ;next jump at address 0400h (A10)
0400 020800  add10:   ljmp add11      ;test address line A10
0800                           .org 0800h      ;next jump at address 0800h (A11)
0800 021000  add11:   ljmp add12      ;test address line A11
1000                           .org 1000h      ;next jump at address 1000h (A12)
1000 022000  add12:   ljmp add13      ;test address line A12
2000                           .org 2000h      ;next jump at address 2000h (A13)
2000 024000  add13:   ljmp add14      ;test address line A13
4000                           .org 4000h      ;last jump at address 4000h (A14)
4000 028000  add14:   ljmp add15      ;test address line A14 and remain
8000                           .org 8000h      ;test address line A15 and remain
                                                ;here
8000 028000  add15:   ljmp add15      ;jump here in a loop
8003                           .end            ;assembler use
```

ADDRESS	MNEMONIC	COMMENT
	.equ ramstart,0000h	;set RAM test start address
	.equ rmstphi,20h	;set RAM test high stop address
	.equ pattern,55h	;determine test pattern
	.equ good,20h	;RAM good pattern P3.5 = 1
	.equ bad,0dfh	;RAM bad pattern, P3.5 = 0
	.org 0000h	;begin test program at 0000h
	mov p3,#0ffh	;set Port 3 high
	mov dptr,#ramstart	;initialize DPTR
test:	mov a,#pattern	;set pattern byte
	movx @dptr,a	;write byte to RAM

Continued

ADDRESS	MNEMONIC	COMMENT
Continued		
	inc dptr	;point to next RAM byte
	mov a,#rmstphi	;check to see if at stop address
	cjne a,dph,test	;if not then loop until done
	mov dptr,#ramstart	;start read–back test
check:	movx a,@dptr	;read byte from RAM
	cjne a,#pattern,fail	;test against what was written
	inc dptr	;go to next byte if tested ok
	mov a,#rmstphi	;check to see if all bytes tested
	cjne a,dph,check	;if not then check again
	mov p3,#good	;checked ok, set Port 3 to good
here:	sjmp here	;stop here
fail:	mov p3,dph	;test failed, get address
	anl p3,#bad	;set 3.5 to zero
	mov p1,dpl	;set Port 1 to low address byte
there:	sjmp there	;stop there
	end	

 COMMENT

Change the ramstart and rmstphi .equ hex numbers to check pages 2000h to 3FFFh, 4000h to 5FFFh, and 6000h to 7FFFh.

Note that a full 16-bit check for end of memory does not have to be done due to page boundaries of (20)00, (40)00, (60)00, and (80)00h.

There is no *halt* command for the 8051; jumps in place serve to perform the halt function.

We have now tested all the external circuitry that has been added to the 8051. The remainder of the chapter is devoted to several subroutines that can be used by the application programs in Chapters 8 and 9.

Timing Subroutines

Subroutines are used by call programs in what is known as a "transparent" manner—that is, the calling program can use the subroutines without being bothered by the details of what is actually going on in the subroutine. Usually, the call program preloads certain locations with data, calls the subroutine, then gets the results back in the preload locations.

The subroutine must take great care to save the values of all memory locations in the system that the subroutine uses to perform internal functions and restore these values before returning to the call program. Failure to save values results in occasional bugs in the main program. The main program assumes that everything is the same both before and after a subroutine is called.

Finally, good documentation is essential so that the user of the subroutine knows precisely how to use it.

Time Delays

Perhaps the most-used subroutine is one that generates a programmable time delay. Time delays may be done by using software loops that essentially do nothing for some period, or by using hardware timers that count internal clock pulses.

The hardware timers may be operated in either a software or a hardware mode. In the software mode, the program inspects the timer overflow flag and jumps when it is set. The hardware mode uses the interrupt structure of the 8051 to generate an interrupt to the program when the timer overflows.

The interrupt method is preferred whenever processor time is scarce. The interrupt mode allows the processor to continue to execute useful code while the time delay is taking place. Both the pure software and timer-software modes tie up the processor while the delay is taking place.

If the interrupt mode is used, then the program *must* have an interrupt handling routine at the dedicated interrupt program vector location specified in Chapter 2. The program must also have programmed the various interrupt control registers. This degree of "nontransparency" generally means that interrupt-driven subroutines are normally written by the user as needed and not used from a purchased library of subroutines.

Pure Software Time Delay

The subroutine named "softime" generates delays ranging from 1 to 65,535 milliseconds by using register R7 to generate the basic 1 millisecond delay. The call program loads the desired delay into registers A (LSB) and B (MSB) before calling Softime.

The key to writing this program is to calculate the exact time each instruction will take at the clock frequency in use. For a crystal of 16 megahertz, each machine cycle (12 clock pulses) is

$$\text{Cycle Time} = \frac{12 \text{ pulses}}{16,000,000 \text{ pulses/s}} = .75 \ \mu s$$

Should the crystal frequency be changed, the subroutine would have to have the internal timing loop number "delay" changed.

Softime

Softime will delay the number of milliseconds expressed by the binary number, from 1 to 65,535d, found in registers A (LSB) and B (MSB). The call program loads the desired delay into registers A and B and calls Softime. Loading zeroes into A and B results in an immediate return.

The number after the comma in the comments section of the following program is the number of cycles for that instruction.

ADDRESS	MNEMONIC	COMMENT
	.equ delay,0ech	;for 996 μs time delay = 222d
softime:	.org 0000h	;set origin
	push 07h	;save R7
	push acc	;save A for A = B = 00 test
	orl a,b	;will be 00 if both 00
	cjne a,#00h,ok	;return if all 00
	pop acc	;keep stack balanced
	sjmp done	
ok:	pop acc	;not all zeroes, proceed
timer:	mov r7,#delay	;initialize R7, 1
onemil:	nop	;tune the loop for 6 cycles, 1
	nop	;this makes 2 cycles total, 1
	nop	;3 cycles total, 1

Continued

ADDRESS	MNEMONIC	COMMENT

Continued

```
                nop                 ;4 cycles total, 1
                djnz r7,onemil      ;count R7 down; 6 cycles total, 2
;
;total delay is 6 cycles (4.5 µs) × 222d = 999d µs.
;
                nop                 ;tune subroutine .75 µs more
;
;total delay is 999.75 µs, which is as close as possible for the
;frequency used (1000 µs = 4000/3 cycles)
;
                djnz acc,timer      ;count A and B down as one
                cjne a,b,bdown      ;A = 00, count B down until = 00
                sjmp done           ;if so then delay is done
bdown:          dec b               ;count B down and time again
                sjmp timer
done:           pop 07h             ;restore R7 to original value
                ret                 ;return to calling routine
                .end
```

── ▷ ── COMMENT ───────────────────────────────────

Note that register A, when used in a defined mnemonic is used as "A." When used as a direct address in a mnemonic (where any add could be used), the equate name ACC is used. The equate usage is also seen for R7, where the name of the register may be used in those mnemonics for which it is specifically defined. For mnemonics that use any add, the actual address must be used.

The restriction on A = B = 00 is due to the fact that the program would initially count A from 00 . . . FFh . . . 00 then exit. If it were desired to be able to use this initial condition for A and B, then an all zero condition could be handled by the test for 0000 used, set a flag for the condition, decrement B from 00 to FFh the first time B is decremented, then reset the flag for the remainder of the program.

The accuracy of the program is poorest for a 1 millisecond delay due to time delay for the rest of the program to set up and return. The actual delay if 0001 is passed to the subroutine is 1014.75 microseconds or an error of 1.5 percent.

Software Polled Timer

A delay that uses the timers to generate the delay and a continuous software flag test (the flag is "polled" to see whether it is set) to determine when the timers have finished the delay is given in this section. The user program signals the total delay desired by passing delay variables in the A and B registers in a manner similar to the pure software delay subroutine. A basic interval of 1 millisecond is again chosen so that the delay may range from 1 to 65,535 ms.

The clock frequency for the timer is the crystal frequency divided by 12, or one machine cycle, which makes each count of the timer .75 microsecond for a 16 megahertz crystal. A 1 millisecond delay gives

$$\text{Count for 1000 microseconds} = 1000/.75 = 1333.33 \ (1333)$$

Due to the fraction, we can not generate a precise 1 millisecond delay using the crystal chosen. If accurate timing is important, then a crystal frequency that is a multiple of 12 must be chosen. Twelve megahertz is an excellent choice for generating accurate time delays, such as for use in systems which maintain a time of day clock.

Timer 0 will be used to count 1333 (0535h) internal clock pulses to generate the basic 1 millisecond delay; registers A and B will be counted down as T0 overflows. The timer counts up, so it will be necessary to put the 2's complement of the desired number in the timer and count up until it overflows.

Timer

The time delay routine named "Timer" uses timer 0 and registers A and B to generate delays from 1 to 65,535d milliseconds. The calling program loads registers A (LSB) and B (MSB) with the desired delay in milliseconds. Loading a delay of 0000h results in an immediate return.

ADDRESS	MNEMONIC	COMMENT
	.equ onemshi,0fah	;2's complement of 535h = FACBh
	.equ onemslo,0cbh	
	.org 0000h	;set program origin
timer:	push tl0	;save timer 0 contents
	push th0	
	cjne a,#00h,go	;test for A = 00
	orl a,b	;A = 00, test for B = 00
	jz done	;A will be 00 if A = B = 00
	clr A	;B is not 00, clear A
go:	anl tcon,#0cfh	;clear timer 0 overflow and run
		;flags in TCON
	anl tmod,#0f0h	;clear T0 part of TMOD, set T0 for
	orl tmod,#01h	;timer operation, mode 1 (16 bit)
onems:	mov tl0,#onemslo	;set T0 to count up from FACBh
	mov th0,#onemshi	
	orl tcon,#10h	;start timer 0
wait:	jbc tf0,dwnab	;poll T0 overflow flag
	sjmp wait	;loop until T0 overflows
dwnab:	anl tcon,#0efh	;stop T0
	djnz acc,onems	;count A down and loop until zero
	cjne a,b,bdown	;if A = B = 00 then done, return
	sjmp done	
bdown:	dec b	;decrement B and count again
	sjmp onems	
done:	pop th0	;restore T0 contents
	pop tl0	
	ret	
	.end	

▷— COMMENT

T0 cannot be used accurately for other timing or counting functions in the user program; thus, there is no need to save the TCON and TMOD bits for T0. T0 itself could be used to store data; it is saved.

This program has no inherent advantage over the pure software delay program; both take up all processor time. The software polled timer has a slight advantage in flexibility in that the

Continued

COMMENT

Continued

number loaded into T0 can be easily changed in the program to shorten or lengthen the basic timing loop. Thus, the call program could also pass the basic timing delay (in other memory locations) and get delays that could be programmed in microseconds or hours.

One way for the program to continue to run while the timer times out is to have the program loop back on itself periodically, checking the timer overflow flag. This looping is the normal operating mode for most programs; if the program execution time is small compared with the desired delay, then the error in the total time delay will be small.

Pure Hardware Delay

If lengthy delays must be done or processor time is so valuable that no time can be wasted for even relatively short software delays, then the time delays must be done using a timer in the interrupt mode. The program given in this section operates in the following manner:

1. The occurrence of a timer overflow will interrupt the processor, which then performs a hardware call to whatever subroutine is located at the dedicated timer flag interrupt address location in ROM.

2. The subroutine determines whether the time delay passed by the using program is finished. (If not, an immediate return is done to the user program at the place where it was interrupted. If the delay is up, then a call to the user part of the program that needed the delay is done, followed by a return to the program where it was interrupted.)

The time delay is initiated by the user program that stores the desired delay at an external RAM location named "Savetime," and then calls "Startime," which sets the timing in motion. The main program then runs while the delay is timing out.

This type of program must use the manufacturer-specified dedicated interrupt locations in ROM that contain the interrupt handling routines. For this reason, the user must have placed some set of instructions at the ROM interrupt location before incorporating the time delay subroutine program in the user program.

In this example, the following three subroutines have been placed at the interrupt location in ROM:

1. Hardtime: a subroutine located at the timer flag interrupt location that determines whether the time delay has expired (If time has not expired, then the subroutine immediately returns to the main user program at the location where it was interrupted by the timer flag; if time is up, then it calls the user program, "Usertime.")

2. Usertime: a subroutine, written by the user, that needed the delay (For this example, the subroutine is simply a return.)

3. Stoptime: a subroutine that stops the timer

■ **Note:** To the assembler, the name of the subroutine can be in *any* combination of *uppercase* or *lowercase:* for example, HARDTIME, hardtime, and HaRdTiMe are all read as the same label name.

The hardware delay subroutine examined here uses timer 1 for the basic delay. When timer 1 overflows and sets the overflow flag, the program will vector to location 001Bh in program memory if the proper bits in the interrupt control registers IE and IP are set.

As in previous examples, the user can set timer 1 for delays of 1 to 65,535 milliseconds by setting the desired delay in external RAM locations "Savetime" (LSB) and "Savetime" + 1 (MSB), which is a two-byte address pointed to by DPTR. Registers A and B cannot be used as in previous examples because to do so would preclude their use for any other purpose in the program.

The hardware delay called "Hardtime" is listed in the following subsection. To avoid confusion as to which is the subroutine and which is the user program, all user code will begin with a label that starts with the name "User." Everything else is the timing routine.

Hardtime

The "Hardtime" subroutine is a hardware-only time delay. To start the delay, IE.7 and IE.3 (EA and ET1) must be set and the subroutine "Startime" called. Three instructions must be assembled at timer 1 location 001Bh: LJMP hardtime, ACALL usertime (with the label "Userdly"), and ACALL stoptime. The priority of the interrupt can be set at bit IP.3 (PT1) to high (1) or low (0). An excerpt from the calling program follows to show these details:

ADDRESS	MNEMONIC	COMMENT
	.equ savetime,0010h	;external RAM address for delay
userpgm:	.org 0000h	;start user program
	sjmp userover	;jump over interrupt addresses
	.org 001bh	;interrupt location for TF1
	ljmp hardtime	;jump to time delay subroutine
userdly:	acall usertime	;called if delay is up
	acall stoptime	;dissable timer interrupt
	reti	;return to main program
userover:	mov dptr,#savetime	;point to delay address
	mov a, #01h	;store desired delay, LSB first
	movx @dptr,a	
	inc dptr	;point to next byte (MSB)
	mov a,#10h	
	movx @dptr,a	;desired delay now stored
	orl ie, #88h	;enable T1 and all interrupts
	acall startime	;start time delay
here:	sjmp here	;loop to simulate user program

```
;
;the user program now continues while timer 1 runs until TF1 = 1.
;the interrupt generated will vector to location 001Bh and execute
;a jump to hardtime that will decrement the contents of savetime
;until the desired time delay has been done; hardtime will return to
;the main program if the delay is not finished, or to userdly if the
;delay is up; userdly returns to call stoptime, which stops the timer
;and returns to the RETI instruction for return to the main program
;
```

startime:	mov thl, #0fah	;set T1 for a 1 ms delay
	mov tll, #0cbh	;(see TIMER example)
	anl tmod,#0fh	;clear T1 part of TMOD
	orl tmod,#40h	;set T1 to timer mode 1
	orl tcon,#40h	;start timer 1
	ret	;return to calling program
hardtime:	push acc	;save registers to be used
	push dph	

Continued

ADDRESS	MNEMONIC	COMMENT
Continued		
	push dpl	
	mov dptr,#savetime	;get pointer to time delay
	movx a,@dptr	;count delay number down to 0000
	dec a	;low byte first
	cjne a,#00h,aff	;check for 0000
	movx @dptr,a	;save low byte = 00
	inc dptr	;get high byte and look for 00
	movx a,@dptr	
	jz done	;done if low, high byte = 0
	sjmp sava	;not 0, delay again
aff:	cjne a,#0ffh,sava	;if low byte = FF dec high
	movx @dptr,a	;save low byte = FF
	inc dptr	;point to high byte
	movx a,@dptr	;count high byte down
	dec a	
	sjmp sava	;save the high byte
done:	pop dpl	;finished, jump to userdly
	pop dph	;restore all registers used
	pop acc	
	ljmp userdly	;continue at user delay
sava:	movx @dptr,a	;delay not up, save byte
	pop dpl	;restore saved registers
	pop dph	
	pop acc	
	acall startime	;start T1 for next 1 ms
	reti	;return to user program

```
;
;the user program ''usertime'' can now be written as needed; a return
;will be used to simulate the user routine.
;
usertime:   ret
;
;after the user program is done then ''stoptime'' will stop timer T1
;and return to the interrupted main program
;
stoptime:   anl tcon,#0bfh      ;stop timer T1
            ret                 ;return to reti
            .end
```

◁— **COMMENT** ————————————————————————————

The minimum usable delay is 1 ms because a 1 ms delay is done to begin the delay interrupt cycle.

All timing routines can be assembled at interrupt location 001Bh if stack space is limited.

The RETI instruction is used when returning to the main program, after each interrupt, while RET instructions are used to return from called routines.

There is no check for an initial delay of 0000h.

Lookup Tables for the 8051

There are many instances in computing when one number must be converted into another number, or a group of numbers, on a one-to-one basis. A common example is to change an ASCII character for the decimal numbers 0 to 9 into the binary equivalent (BCD) of those numbers. ASCII 30h is used to represent 00d, 31h is 01d, and so on, until ASCII 39h is used for 09d.

Clearly, one way to convert from ASCII to BCD is to subtract a 30h from the ASCII character. Another approach uses a table in ROM that contains the BCD numbers 00 to 09. The table is stored in ROM at addresses that are related to the ASCII character that is to be converted to BCD. The ASCII character is used to form part of the address where its equivalent BCD number is stored. The contents of the address "pointed" to by the ASCII character are then moved to a register in the 8051 for further use. The ASCII character is then said to have "looked up" its equivalent BCD number.

For example, using ASCII characters 30h to 39h we can construct the following program, at the addresses indicated, using .db commands:

ADDRESS	MNEMONIC	COMMENT
	.org 1030h	;start table at ROM location 1030h
	.db 00h	;location 1030h contains 00 BCD
	.db 01h	;location 1031h contains 01 BCD
	.db 02h	;location 1032h contains 02 BCD
	.db 03h	;location 1033h contains 03 BCD
	.db 04h	;location 1034h contains 04 BCD
	.db 05h	;location 1035h contains 05 BCD
	.db 06h	;location 1036h contains 06 BCD
	.db 07h	;location 1037h contains 07 BCD
	.db 08h	;location 1038h contains 08 BCD
	.db 09h	;location 1039h contains 09 BCD

Each address whose low byte is the ASCII byte contains the BCD equivalent of that ASCII byte. If the DPTR is loaded with 1000h and A is loaded with the desired ASCII byte, then a MOVC A,@A+DPTR will move the equivalent BCD byte for the ASCII byte in A to A.

Lookup tables may be used to perform very complicated data translation feats, including trigonometric and exponential conversions. While lookup tables require space in ROM, they enable conversions to be done very quickly, far faster than using computational methods.

The 8051 is equipped with a set of instructions that facilitate the construction and use of lookup tables: the MOVC A,@A+DPTR and the MOVC A,@A+PC. In both cases A holds the pointer, or some number calculated from the pointer, which is also called an "offset." DPTR or PC holds a "base" address that allows the data table to be placed at any convenient location in ROM. In the ASCII example just illustrated, the base address is 1000h, and A holds an offset number ranging from 30h to 39h.

Typically, PC is used for small "local" tables of data that may be included in the body of the program. DPTR might be used to point to large tables that are normally assembled at the end of program code.

In both cases, the desired byte of data is found at the address in ROM that is equal to base + offset. Figure 7.6 demonstrates how the final address in the lookup table is calculated using the two base registers.

FIGURE 7.6 MOVC ROM Address Calculations

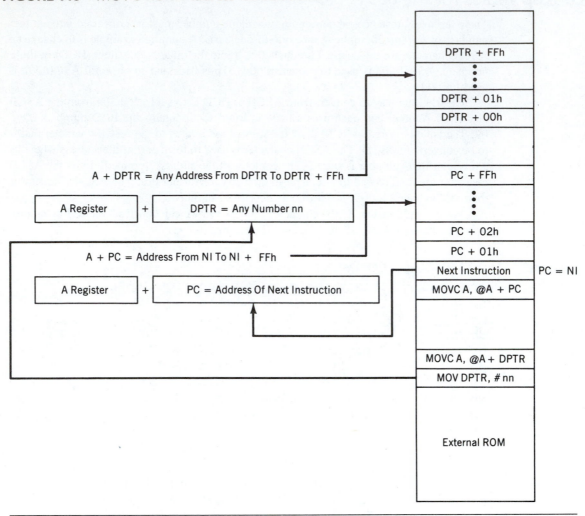

One limitation of lookup tables might be the appearance that only 256 different values—corresponding to the 256 different values that A might hold—may be put in a table. This limitation can be overcome by using techniques to alter the DPTR such that the base address is changed in increments of 256 bytes. The same offset in A can point to any number of data bytes in tables that differ only by the beginning address of the base. For example, by changing the number loaded in DPTR from 1000h to 1100h in the ASCII-to-BCD table given previously, the ASCII byte in A can now point to an entirely new set of conversion bytes.

Both PC and DPTR base address programs are given in the examples that follow.

PC as a Base Address

Suppose that the number in A is known to be between 00h and 0Fh and that the number in A is to be squared. A could be loaded into B and a MUL AB done or a local lookup table constructed.

The table cannot be placed directly after the MOVC instruction. A jump instruction must be placed between the MOVC and the table, or the program soon fetches the first data byte of the table and executes it as code. Remember also that the PC contains the address of the jump instruction (the Next Instruction, after the MOVC command) when the table address is computed.

Pclook

The program "pclook" looks up data in a table that has a base address in the PC and the offset in A. After the MOVC instruction, A contains the number that is the square of the original number in A.

```
ADDRESS      MNEMONIC         COMMENT
             .org 0000h
pclook:      mov a,#0ah        ;find the square of 0Ah (64h)
             add a,#02h        ;adjust for two byte sjmp over
             movc a,@a+pc      ;get equivalent data from table to A
             sjmp over         ;jump over the lookup table
;
;the lookup table is inserted here, at PC + 2. (PC = 0005h)
;
             .db 00h           ;begin table here, 00^2 = 00
             .db 01h           ;01^2 = 01d
             .db 04h           ;02^2 = 04d
             .db 09h           ;03^2 = 09d
             .db 10h           ;04^2 = 16d
             .db 19h           ;05^2 = 25d
             .db 24h           ;06^2 = 36d
             .db 31h           ;07^2 = 49d
             .db 40h           ;08^2 = 64d
             .db 51h           ;09^2 = 81d
             .db 64h           ;0A^2 = 100d
             .db 79h           ;0B^2 = 121d
             .db 90h           ;0C^2 = 144d
             .db 0a9h          ;0D^2 = 169d
             .db 0c4h          ;0E^2 = 196d
             .db 0e1h          ;0F^2 = 225d
over:        sjmp over         ;simulate rest of user program
             .end
```

Figure 7.7 shows the assembled listing of this program and the resulting address of the table relative to the MOVC instruction.

---▷ COMMENT ───

The number added to A reflects the number of bytes in the SJMP instruction. If more code is inserted between the MOVC and the table, a similar number of bytes must be added. Adding bytes can result in overflowing A when the sum of these adjusting bytes and the contents of A exceed 255d. If this happens, the lookup data must be limited to the number of bytes found by subtracting the number of adjustment bytes from 255d.

FIGURE 7.7 Lookup Table using the PC

```
0000                    .org 0000h
0000 740A pclook:    mov a,#0ah          ;find the square of 0Ah (64h)
0002 2402            add a,#02h          ;adjust for two byte sjmp over
0004 83              movc a,@a+pc        ;get equivalent data from table
                                         ;to A
0005 8010            sjmp over           ;jump over the lookup table
0007          ;the lookup table is inserted here, at PC + 2 (PC = 0005h)
0007 00              .db 00h             ;begin table here, 00^2 = 00
0008 01              .db 01h             ;01^2 = 01d
0009 04              .db 04h             ;02^2 = 04d
000A 09              .db 09h             ;03^2 = 09d
000B 10              .db 10h             ;04^2 = 16d
000C 19              .db 19h             ;05^2 = 25d
000D 24              .db 24h             ;06^2 = 36d
000E 31              .db 31h             ;07^2 = 49d
000F 40              .db 40h             ;08^2 = 64d
0010 51              .db 51h             ;09^2 = 81d
0011 64              .db 64h             ;0A^2 = 100d
0012 79              .db 79h             ;0B^2 = 121d
0013 90              .db 90h             ;0C^2 = 144d
0014 A9              .db 0A9h            ;0D^2 = 169d
0015 C4              .db 0C4h            ;0E^2 = 196d
0016 E1              .db 0E1h            ;0F^2 = 225d
0017 80F  over:      sjmp over           ;simulate rest of user program
0019                 .end
```

DPTR as a Base Address

The DPTR is used to construct a lookup table in the next example. Remove the restriction that the number in A must be less than 10h and let A hold any number from 00h to FFh. The square of any number larger than 0Fh results in a four-byte result; store the result in registers R0 (LSB) and R1 (MSB).

Two tables are constructed in this section: one for the LSB and the second for the MSB. A points to both bytes in the two tables, and the DPTR is used to hold two base addresses for the two tables. The entire set of two tables, each with 256 entries, will not be constructed for this example. The beginning and example values are shown as a skeleton of the entire table.

Dplook

The lookup table program "dplook" holds the square of any number found in the A register. The result is placed in R0 (LSB) and R1 (MSB). A is stored temporarily in R1 in order to point to the MSB byte.

ADDRESS	MNEMONIC	COMMENT
	.equ lowbyte,0200h	;base address of LSB table
	.equ hibyte,0300h	;base address of MSB table
	.org 0000h	

Continued

ADDRESS	MNEMONIC	COMMENT
dplook:	mov a,#5ah	;find the square of 5Ah (1FA4h)
	mov rl,a	;store A for later use
	mov dptr,#lowbyte	;set DPTR to base address of LSB
	movc a,@a+dptr	;get LSB
	mov r0,a	;store LSB in R0
	mov a,rl	;recover A for pointing to MSB
	mov dptr,#hibyte	;set DPTR to base address of MSB
	movc a,@a+dptr	;get MSB
	mov rl,a	;store MSB in Rl
here:	sjmp here	;simulate rest of user program
	.org lowbyte	;place LSB table starting here
	.db 00h	;00^2 = 0000
	.db 0lh	;0l^2 = 0001

```
;place rest of table up to the LSB of 59^2 here
        .org lowbyte + 5ah    ;put LSB of 5A^2 here
        .db 0a4h              ;LSB is A4h
;place rest of LSB table here
        .org hibyte           ;place MSB table starting here
        .db 00h               ;00^2 = 0000
        .db 00h               ;0l^2 = 0001
;place rest of table up to the MSB of 59^2 here
        .org hibyte + 5ah     ;put MSB of 5A^2 here
        .db 1fh               ;MSB is 1Fh
;place rest of MSB table here
        .end
```

▷ **COMMENT** ─────────────────────────────────────

Note that there are no jumps to "get over" the tables; the tables are normally placed at the end of the program code.

A does not require adjustment; DPTR is a constant.

Figure 7.8 shows the assembled code; location 025Ah holds the LSB of 5A^2, and location 035Ah holds the MSB.

Serial Data Transmission

The hallmark of contemporary industrial computing is the linking together of multiple processors to form a "local area network" or LAN. The degree of complexity of the LAN may be as simple as a microcontroller interchanging data with an I/O device, as complicated as linking multiple processors in an automated robotic manufacturing cell, or as truly complex as the linking of many computers in a very high speed, distributed system with shared disk and I/O resources.

All of these levels of increasing sophistication have one feature in common: the need to send and receive data from one location to another. The most cost-effective way to meet this need is to send the data as a serial stream of bits in order to reduce the cost (and bulk) of multiple conductor cable. Optical fiber bundles, which are physically small, can be used for parallel data transmission. However, the cost incurred for the fibers, the terminations, and the optical interface to the computer currently prohibit optical fiber use, except in those cases where speed is more important than economics.

FIGURE 7.8 Lookup Table using the DPTR

```
0200                        .equ lowbyte,0200h      ;base address of LSB table
0300                        .equ hibyte,0300h       ;base address of MSB table
0000                        .org 0000h
0000 745A dplook:           mov a,#5ah              ;find the square of 5Ah
                                                    ;(1FA4h)
0002 F9                     mov r1,a                ;store A for later use
0003 900200                 mov dptr,#lowbyte       ;set DPTR to base address
                                                    ;of LSB
0006 93                     movc a,@a+dptr          ;get LSB
0007 F8                     mov r0,a                ;store LSB in R0
0008 E9                     mov a,r1                ;recover A for pointing
                                                    ;to MSB
0009 900300                 mov dptr,#hibyte        ;set DPTR to base address
                                                    ;of MSB
000C 93                     movc a,@a+dptr          ;get MSB
000D F9                     mov r1,a                ;store MSB in R1
000E 80FE here:             sjmp here               ;simulate rest of user
                                                    ;program
0200                        .org lowbyte            ;place LSB table starting
                                                    ;here
0200 00                     .db 00h                 ;00^2 = 0000
0201 01                     .db 01h                 ;01^2 = 0001
0202          ;place rest of table up to the LSB of 59^2 here
025A                        .org lowbyte + 5ah      ;put LSB of 5A^2 here
025A A4                     .db 0a4h                ;LSB is A4h
025B          ;place rest of LSB table here
0300                        .org hibyte             ;place MSB table starting
                                                    ;here
0300 00                     .db 00h                 ;00^2 = 0000
0301 00                     .db 00h                 ;01^2 = 0001
0302          ;place rest of table up to the MSB of 59^2 here
035A                        .org hibyte + 5ah       ;put MSB of 5A^2 here
035A 1F                     .db 1fh                 ;MSB is 1Fh
035B          ;place rest of MSB table here
035B                        .end
```

So pervasive is serial data transmission that special integrated circuits, dedicated solely to serial data transmission and reception, appeared commercially in the early 1970s. These chips, commonly called "universal asynchronous receiver transmitters," or UARTS, perform all the serial data transmission and reception timing tasks of the most popular data communication scheme still in use today: serial 8-bit ASCII coded characters at predefined bit rates of 300 to 19200 bits per second.

Asynchronous transmission utilizes a start bit and one or more stop bits, as shown in Figure 7.9, to alert the receiving unit that a character is about to arrive and to signal the end of a character. This "overhead" of extra bits, with the attendant slowing of data byte rates, has encouraged the development of synchronous data transmission schemes. Synchronous data transmission involves alerting the receiving unit to the arrival of data

FIGURE 7.9   Asynchronous 8-Bit Character

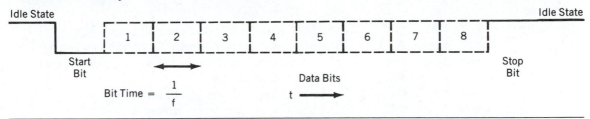

by a unique pattern that starts data transmission, followed by a long string of characters. The end of transmission is signaled by another unique pattern, usually containing error-checking characters.

Each scheme has its advantages. For relatively short or infrequent messages, the asynchronous mode is best; for long messages or constant data transmission, the synchronous mode is superior.

The 8051 contains serial data transmission/receiver circuitry that can be programmed to use four asynchronous data communication modes numbered from 0 to 3. One of these, mode 1, is the standard UART mode, and three simple asynchronous communication programs using this mode will be developed here. More complicated asynchronous programs that use all of the communication modes will be written in Chapter 9.

Character Transmission Using a Time Delay

Often data transmission is unidirectional from the microcontroller to an output device, such as a display or a printer. Each character sent to the output device takes from 33.3 to .5 milliseconds to transmit, depending upon the baud rate chosen. The program must wait until one character is sent before loading the next, or data will be lost. A simple way to prevent data loss is to use a time delay that delays the known transmission time of one character before the next is sent.

Sendchar

A program called "Sendchar" takes the character in the A register, transmits it, delays for the transmission time, and then returns to the calling program. Timer 1 must be used to set the baud rate, which is 1200 baud in this example. The delay for one ten-bit character is 1000/120 or 8.4 milliseconds. The software delay developed in Section 7.5 is used for the delay with the basic delay period of 1 milliseconds changed to .1 milliseconds by redefining "delay." Timer 1 needs to generate a final baud rate of 1200 at SBUF. Using a 16 megahertz crystal, the reload number is $256 - 16E6/(16 \times 12 \times 1200)$, which is 186.6 or integer 187. This yields an actual rate of 1208.

ADDRESS	MNEMONIC	COMMENT
	.org 0000h	
	.set delay,16h	;basic delay = 22d × 4.5 = 99 μs
	mov a,#'A'	;for this example, send an A
	acall sendchar	;send it
here:	sjmp here	;simulate rest of user program
;		

Continued

ADDRESS	MNEMONIC	COMMENT

Continued

```
sendchar:   anl  tmod,#0fh      ;alter timer 1 configuration only
            orl  tmod,#20h      ;set timer 1 for mode 2 (auto reload)
            mov  thl,#0bbh      ;set reload number to 187d (256 - 69)
            orl  pcon,#80h      ;set SMOD bit to 1
            orl  tcon,#40h      ;start timer 1 by setting TR1
            mov  scon,#40h      ;set serial port to mode 1
            mov  sbuf,a         ;load transmit register and wait
            mov  a,#54h         ;delay for 8.4 ms (84d = 54h)
            acall softime       ;wait
            ret                 ;character now sent
;
;softime will be simulated by a return instruction
softime:    ret
            .end                ;assembler use only
```

▷ COMMENT ─────────────────────────────────────

If timer 1 and the serial port have different uses in the user program, then push and pop affected control registers. But remember, T1 and SBUF can only be used for one function at any given time.

The use of the .set statement lets the user change the basic delay interval to different values in the same program.

The 16 megahertz crystal does not yield convenient standard baud rates of 300, 1200, 2400, 4800, 9600, or 19200. The errors using this crystal for these rates are given in the following table:

RATE	ERROR (%)
300	.08
1200	.64
4800	2.12
9600	3.55
19200	8.51

The error grows for higher baud rates as ever smaller reload numbers are rounded to the nearest integer. Using an 11.059 megahertz crystal reduces the errors to less than .002 percent at the cost of speed of program execution.

Character Transmission by Polling

An alternative to waiting a set time for transmission is to monitor the TI flag in the SCON register until it is set by the transmission of the last character written to SBUF. The polling routine must reset TI before returning to the call program. Failure to reset TI will inhibit all calls after the first, stopping all data transmission except the first character.

This technique has the advantage of simplicity; less code is used, and the routine does not care what the actual baud rate is. In this example, it is assumed that the timer 1 baud rate has been established at the beginning of the program in a manner similar to that used in the previous example.

Xmit

The subroutine "xmit" polls the TI flag in the SCON register to determine when SBUF is ready for the next character. The calling part of the user program follows:

ADDRESS	MNEMONIC	COMMENT
	.org 0000h	
	mov a,#'3'	;send an ASCII 3 for this example
	acall xmit	;send the character using xmit
here:	sjmp here	;simulate remainder of user program
;		
xmit:	mov sbuf,a	;transmit the contents of A and wait
wait:	jnb scon.1,wait	;loop until TI = 1 (SBUF is empty)
	clr scon.1	;reset TI to 0
	ret	
	.end	

 **COMMENT**

TI remains a 0 until SBUF is empty; when the 8051 is reset, or upon power up, TI is set to 0.

Interrupt-Driven Character Transmission

The third method of determining when transmission is finished is to use the interrupt structure of the 8051. One interrupt vector address in program code, location 0023h, is assigned to both the transmit interrupt, TI, and the receive interrupt, RI. When a serial interrupt occurs, a hardware call to location 0023h accesses the interrupt handling routine placed there by the programmer.

The user program "calls" the subroutine by loading the character to be sent into SBUF and enabling the serial interrupt bit in the EI register. The user program can then continue executing. When SBUF becomes empty, TI will be set, resulting in an immediate vector to 0023h and the subroutine placed there executed. The subroutine at 0023h, called "serial," will reset TI and then return to the user program at the place where it was interrupted.

This scheme is satisfactory for testing the microprocessor when only one character is sent from the program. Long strings of character transmission will overload SBUF. Chapter 9 contains routines that will build on this technique and send arbitrarily long strings with no loss of data.

SBUFR

An interrupt-driven data transmission routine for one character which is assembled at the interrupt vector location 0023h. A portion of the user program that activates the interrupt routine is shown.

ADDRESS	MNEMONIC	COMMENT
	.org 0000h	
sbufr:	sjmp user	;jump over interrupt vectors
	.org 0023h	;put serial interrupt routine here

Continued

ADDRESS	MNEMONIC	COMMENT
Continued		
serial:	clr scon.1	;clear TI
	reti	;return to interrupted user program
;		
user:	mov sbuf,#'X'	;send an X in this example
	orl ie,#90h	;enable serial interrupt
here:	sjmp here	;simulate remainder of program
	.end	

 COMMENT

If TI is not cleared before the RETI instruction is used, there will be an immediate interrupt and vector back to 0023h.

RETI is used to reset the entire interrupt structure, not to clear any interrupt bits.

Receiving Serial Data

Transmissions from outside sources to the 8051 are not predictable unless an elaborate time-of-day clock is maintained at the sender and receiver. Messages can then be sent at predefined times. A time-of-day clock generally ties up timers at both ends to generate the required "wake-up" calls.

Two methods are normally used to alert the receiving program that serial data has arrived: software polling or interrupt driven. The sending entity, or "talker," transmits data at random times, but uses an agreed-upon baud rate and data transmission mode. The receiving unit, commonly dubbed the "listener," configures the serial port to the mode and baud rate to be used and then proceeds with its program.

If one programmer were responsible for the talker and another for the listener, lively discussions would ensue when the units are connected and data interchange does not take place. One common method used to test communication programs is for each programmer to use a terminal to simulate the other unit. When the units are connected for the final test, a CRT terminal in a transparent mode, which shows all data transmitted in both directions, is connected between the two systems to show what is taking place in the communication link.

Polling for Received Data

Polling involves periodically testing the received data flag RI and calling the data receiving subroutine when it is set. Care must be taken to remember to reset RI, or the same character will be read again. Reading SBUF does *not* clear the data in SBUF or the RI flag.

The program can sit in a loop, constantly testing the flag until data is received, or run through the entire program in a circular manner, testing the flag on each circuit of the program. The loop approach guarantees that the data be read as soon as it is received; however, very little else will be accomplished by the program while waiting for the data. The circular approach lets the program run while awaiting the data.

In order not to miss any data, the circular approach requires that the program be able to run a complete circuit in the time it takes to receive one data character. The time restraint on the program is not as stringent a requirement as it may first appear. The receiver is double buffered, which lets the reception of a second character begin while a previous character remains unread in SBUF. If the first character is read before the last bit of the second is complete, then no data will be lost. This means that, after a two-character burst, the program still must finish in one-character time to catch a third.

The character time is the number of bits per character divided by the baud rate. For serial data transmission mode 1, a character uses ten bits: start, eight code bits, and stop. A 1200 baud rate, which might be typical for a system where the talker and listener do not interchange volumes of data, results in a character rate of 120 characters per second, or a character time of 8.33 milliseconds. Using an average of 18 oscillator periods per instruction, each instruction will require 1.13 microseconds to execute, enabling a program length of 7371 instructions. This large machine language program will suffice for many simple control and monitoring applications where data transmission rates are low. If more time is needed, the baud rate could be reduced to as low as 300 baud, yielding a program size of over 29K bytes, which approaches half the maximum size of the ROM in our example 8051 design.

The polling program for the loop approach follows:

ADDRESS	MNEMONIC	COMMENT
here:	jnb scon.0,here	;wait here until RI = 1
	clr scon.0	;clear the RI bit
	acall getchar	;getchar is some user routine
		;which reads SBUF

The circular approach is very similar:

ADDRESS	MNEMONIC	COMMENT
	jnb scon.0 there	;test for RI = 1, go on if not
	clr scon.0	;clear the RI bit
	acall getchar	;call user routine
there:	sjmp there	;rest of user program getchar:
	ret	;simulate user routine
	.end	

Interrupt-Driven Data Reception

When large volumes of data must be received, the data rate will overwhelm the polling approach unless the user program is extremely short, a feature not usually found in systems in which large amounts of data are interchanged. Interrupt-driven systems allow the program to run with brief pauses to read the received data. In Chapter 9, a program is developed that allows for the reception of long strings of data in a manner completely transparent to the user program.

Intdat

This interrupt-driven data reception subroutine assembles the program at 0023h, which is the serial interrupt vector location.

ADDRESS	MNEMONIC	COMMENT
	.org 0000h	
intdat:	orl ie,#90h	;enable serial and all interrupts
	sjmp over	;jump over the interrupt locations

Continued

ADDRESS	MNEMONIC	COMMENT
Continued		
	.org 0023h	;put serial interrupt program here
	jbc scon.1,xmit	;if TI bit set, clear it and jump
	clr scon.0	;must have been RI, clear it
	lcall recv	;call receive subroutine
	reti	;return to program where interrupted
xmit:	lcall trans	;call transmit program
	reti	;return to program where interrupted
over:	sjmp over	
trans:	ret	;dummy transmit/receive routines
recv:	ret	

 COMMENT ─────────────────────────────

If both RI and TI are set, this routine will service the transmit function first. After the RETI, which follows the LCALL to trans, the RI bit will still be set, causing an immediate interrupt back to location 0023h where the receive routine will be called.

If the transmit or receive subroutines that are called take longer to execute than the character time, then data will be lost. Long subroutine times would be highly unusual; however, it is possible to overload any system by constant data reception.

Summary

An 8051 based microprocessor system has been designed that incorporates many features found in commercial designs. The design can be easily duplicated by the reader and uses external EPROM and RAM so that test programs may be exercised. Various size memories may be used by the impecunious to reduce system cost.

The design features are

External RAM: 8K to 32K bytes

External ROM: 8K to 64K bytes

I/O ports: 1–8 bit, port 1

Other ports: port 3.0 (RXD)
3.1 (TXD)
3.2 ($\overline{\text{INT0}}$)
3.3 ($\overline{\text{INT1}}$)
3.4 (T0)
3.5 (T1)

Crystal: 16 megahertz

Other crystal frequencies may be used to generate convenient timing frequencies. The design can be modified to include a single step capability (see Problem 2).

Methods of adding additional ports to the basic design are discussed and several example circuits that indicate the expansion possibilities of the 8051 are presented.

Programs written to test the design can be used to verify any prototypes that are built by the reader. These tests involve verifying the proper operation of the ROM and RAM connections.

Several programs and subroutines are developed that let the user begin to exercise the 8051 instruction code and hardware capabilities. This code can be run on the simulator or on an actual prototype. These programs cover the most common types found in most applications:

Time delays: software; timer, software polled; timer, interrupt driven

Lookup Tables: PC base, DPTR base

Serial data communications transmission: time delay, software polled, interrupt driven

Serial data communications reception: software polled, interrupt driven

The foundations laid in this chapter will be built upon by example application programs and hardware configurations found in Chapters 8 and 9.

Problems

1. Determine whether the 8051 can be made to execute a single program instruction (single-stepped) using external circuitry (no software) only.

2. Outline a scheme for single-stepping the 8051 using a combination of hardware and software. (Hint: use an $\overline{\text{INTX}}$.)

3. While running the EPROM test, it is found that the program cannot jump from 2000h to 4000h successfully. Determine what address line(s) is faulty.

4. Calculate the error for the delay program "Softime" when values of 2d, 10d and 1000d milliseconds are passed in A and B.

5. The program "Softime" has a bug. When A = 00h the delay becomes: $(B+1)d \times 256d \times$ delay. Find the bug and fix it without introducing a new bug.

6. Find the shortest and longest delays possible using "Softime" by changing only the equate value of the variable "delay."

7. Give a general description of how you would test any time delay program. (Hint: use a port pin.)

8. In the discussion for the program named "Timer," the statement is made that an accurate 1 ms delay cannot be done due to the need for a count of 1333.33 using a 16 megahertz clock. Find a way to generate an accurate 60 second delay using T0 for the basic delay and some registers to count the T0 overflows.

9. Calculate the shortest and longest delays possible using the program named "Timer" by changing the initial value of T0.

10. Why is there no check for an initial timing value of 0000h in the program named "Hardtime"?

11. Write a lookup table program, using the PC as the base, that finds a one-byte square root (to the nearest whole integer) of any number placed in A. For example, the square roots of 01 and 02 are both 01, while the roots of 03 and 04 are 02. Calculate the first four and last four table values.

12. Write a lookup table, using the DPTR as the base, that finds a two-byte square root of the number in A. The first byte is the integer value of the root, and the second byte is the fractional value. For example, the square root of 02 is 01.6Ah. Calculate four first and last table values.

13. Write a lookup table program that converts the hex number in A (0–F) to its ASCII equivalent.

14. A PC based lookup table, which contains 256d values, is placed 50h bytes after the MOVC instruction that accesses it. Construct the table, showing where the byte associated with A = 00h is located. Find the largest number which can be placed in A to access the table.

15. Construct a lookup table program that converts the hex number in A to an equivalent BCD number in registers R4 (MSB) and R5 (LSB).

16. Reverse Problem 15 and write a lookup table program that takes the BCD number in R4 (MSB) and R5 (LSB) and converts it to a hex number in A.

17. Verify the errors listed for the 16 megahertz crystal in the third comment after the program named "Sendchar."

18. Verify the error listed for the 11.059 megahertz crystal in the fourth comment after the program named "Sendchar."

19. Does asynchronous communication between two microprocessors have to be done at standard baud rates? Name one reason why you might wish to use standard rates.

20. Write a test program that will "loop test" the serial port. The output of the serial port (TXD) is connected to the input (RXD), and the test program is run. Success is indicated by port 1 pin 1 going high.

21. What is the significance of the transmit flag, TI, when it is cleared to 0? When set to 1?

22. Using the programmable port of Figure 7.3, write a program that will configure all ports as outputs, and write a 55h to each.

23. Repeat problem 22 using the memory-mapped programmable port of Figure 7.4.

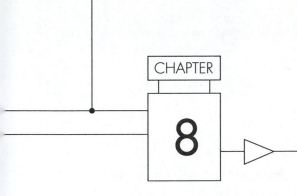

CHAPTER

8

Applications

Chapter Outline

Introduction
Keyboards
Displays
Pulse Measurement

Multiple Interrupts
Putting it all Together
Summary

Introduction

Microcontrollers tend to be underutilized in many applications. There are several reasons for this anomaly. Principally, the devices are so inexpensive that it makes little economic sense to try to select an optimal device for each application. A new microcontroller involves the expense of new development software and training for the designers and programmers that could easily cost more than the part savings. Also, some members of the technical community are unfamiliar with the microcontroller due to a dearth of established academic course offerings on the subject. These individuals tend to apply classic eight-bit microprocessor families to problems that are more economically served by a microcontroller. Finally, there is always the pressure to use the latest multibyte processor for marketing reasons or just to keep up with the "state of the art."

The result of this application pattern is that microcontrollers tend to become obsolete at a slower rate than their CPU cousins. The microcontroller will absorb more eight-bit CPU applications as the economic advantage of using microcontrollers becomes compelling.

Application examples in a textbook present a picture of use that supports the previously-made claim of underutilization. Limitations on space, time, and the patience of the reader preclude the inclusion of involved, multi-thousand line, real-time examples. We will, instead, look at pieces of larger problems, each piece representing a task commonly found in most applications.

One of the best ways to get a "feel" for a new processor is to examine circuits and programs that address easily visualized applications and then to write variations. To assist in this process, we will study in detail the following typical hardware configurations and their accompanying programs:

Keyboards

Displays

Pulse measurements

A/D and D/A conversions

Multi-source interrupts

The hardware and software are inexorably linked in the examples in this chapter. The choice of the first leads to the programming techniques of the second. The circuit designer should have a good understanding of the software limitations faced by the programmer. The programmer should avoid the temptation of having all the tricky problems handled by the hardware.

Keyboards

The predominant interface between humans and computers is the keyboard. These range in complexity from the "up-down" buttons used for elevators to the personal computer QWERTY layout, with the addition of function keys and numeric keypads. One of the first mass uses for the microcontroller was to interface between the keyboard and the main processor in personal computers. Industrial and commercial applications fall somewhere in between these extremes, using layouts that might feature from six to twenty keys.

The one constant in all keyboard applications is the need to accommodate the human user. Human beings can be irritable. They have little tolerance for machine failure; watch what happens when the product isn't ejected from the vending machine. Sometimes they are bored, or even hostile, towards the machine. The hardware designer has to select keys that will survive in the intended environment. The programmer must write code that will anticipate and defeat inadvertent and also deliberate attempts by the human to confuse the program. It is very important to give instant feedback to the user that the key hit has been acknowledged by the program. By the light a light, beep a beep, display the key hit, or whatever, the human user must know that the key has been recognized. Even feedback sometimes is not enough; note the behavior of people at an elevator. Even if the "up" light is lit when we arrive, we will push it again to let the machine know that "I'm here too."

Human Factors

The keyboard application program must guard against the following possibilities:

More than one key pressed (simultaneously or released in any sequence)

Key pressed and held

Rapid key press and release

All of these situations can be addressed by hardware or software means; software, which is the most cost effective, is emphasized here.

Key Switch Factors

The universal key characteristic is the ability to bounce: The key contacts vibrate open and close for a number of milliseconds when the key is hit and often when it is released. These rapid pulses are not discernable to the human, but they last a relative eternity in the microsecond-dominated life of the microcontroller. Keys may be purchased that do not bounce, keys may be debounced with RS flip-flops, or debounced in software with time delays.

Keyboard Configurations

Keyboards are commercially produced in one of the three general hypothetical wiring configurations for a 16-key layout shown in Figure 8.1. The lead-per-key configuration is typically used when there are very few keys to be sensed. Since each key could tie up a port pin, it is suggested that the number be kept to 16 or fewer for this keyboard type. This configuration is the most cost effective for a small number of keys.

The X–Y matrix connections shown in Figure 8.1 are very popular when the number of keys exceeds ten. The matrix is most efficient when arranged as a square so that N leads for X and N leads for Y can be used to sense as many as N^2 keys. Matrices are the most cost effective for large numbers of keys.

FIGURE 8.1 Hypothetical Keyboard Wiring Configurations

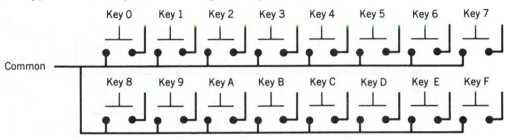

(a) Lead - Per - Key Keyboard

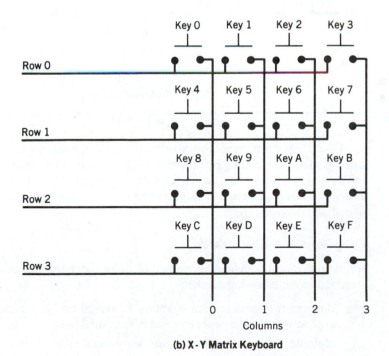

(b) X - Y Matrix Keyboard

Continued

FIGURE 8.1 Continued

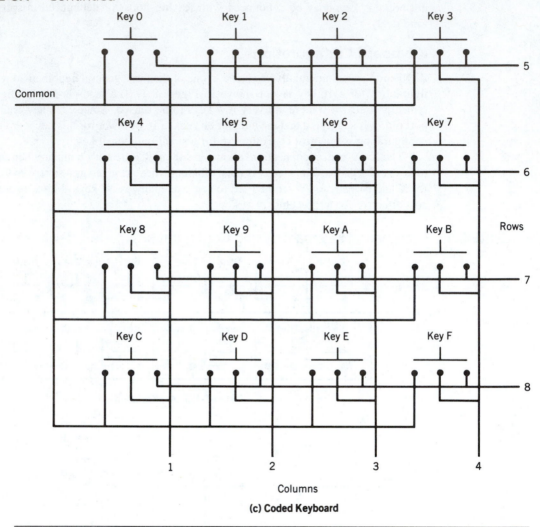

(c) **Coded Keyboard**

Coded keyboards were evolved originally for telephonic applications involving touch-tone signaling. The coding permits multiple key presses to be easily detected. The quality and durability of these keypads are excellent due to the high production volumes and intended use. They are generally limited to 16 keys or fewer, and tend to be the most expensive of all keyboard types.

Programs for Keyboards

Programs that deal with humans via keyboards approach the human and keyswitch factors identified in the following manner:

Bounce: A time delay that is known to exceed the manufacturer's specification is used to wait out the bounce period in both directions.

Multiple keys: Only patterns that are generated by a valid key pressed are accepted—all others are ignored—and the first valid pattern is accepted.

Key held: Valid key pattern accepted after valid debounce delay; no additional keys accepted until all keys are seen to be up for a certain period of time.

Rapid key hit: The design is such that the keys are scanned at a rate faster than any human reaction time.

The last item brings up an important point: Should the keyboard be read as the program loops (software polled) or read only when a key has been hit (interrupt driven)?

In general, the smaller keyboards (lead-per-key and coded) can be handled either way. The common lead can be grounded and the key pattern read periodically. Or, the lows from each can be active-low ORed, as shown in Figure 8.2, and connected to one of the external $\overline{\text{INTX}}$ pins.

Matrix keyboards are scanned by bringing each X row low in sequence and detecting a Y column low to identify each key in the matrix. X–Y scanning can be done by using dedicated keyboard scanning circuitry or by using the microcontroller ports under program control. The scanning circuitry adds cost to the system. The programming approach takes processor time, and the possibility exists that response to the user may be sluggish if the program is busy elsewhere when a key is hit. Note how long your personal computer takes to respond to a break key when it is executing a print command, for instance. The choice between adding scanning hardware or program software is decided by how busy the processor is and the volume of entries by the user.

FIGURE 8.2 Lead-per-Key and Coded Keyboard Interrupt Circuits

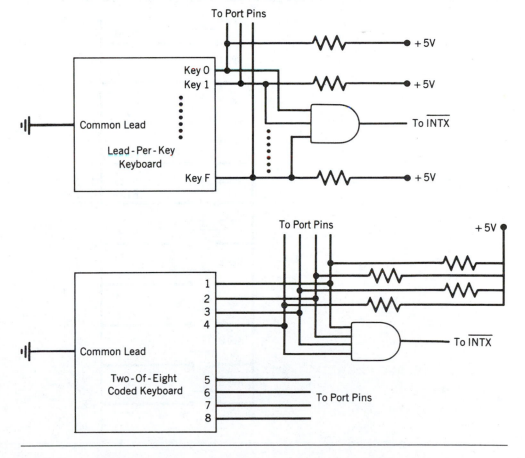

A Scanning Program for Small Keyboards

Assume that a lead-per-key keyboard is to be interfaced to the microcontroller. The keyboard has ten keys (0–9), and the debounce time, when a key is pressed or released, is 20 milliseconds. The keyboard is used to select snacks from a vending machine, so the processor is only occupied when a selection is made. The program constantly scans the keyboard waiting for a key to be pressed before calling the vending machine actuator subroutine. The keys are connected to port 1 (0–7) and ports 3.2 and 3.3 (8–9), as shown in Figure 8.3.

The 8031 works best when handling data in byte-sized packages. To save internal space, the ten-bit word representing the port pin configuration is converted to a single-byte number.

Because the processor has nothing to do until the key has been detected, the time delay "Softime" (see Chapter 7) is used to debounce the keys.

Getkey

The routine "Getkey" constantly scans a ten-key pad via ports 0 and 3. The keys are debounced in both directions and an "all-up" period of 50 milliseconds must be seen before a new key will be accepted. Invalid key patterns (more than one port pin low) are rejected.

FIGURE 8.3 Keyboard Configuration for "Getkey" and "Inkey" Programs

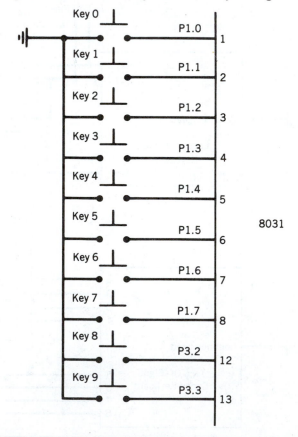

ADDRESS	MNEMONIC	COMMENT
	.equ bounce,14h	;set debounce delay to 20d ms
	.equ next,32h	;set interval between keys to ;50d ms
	.equ newkey,70h	;store accepted key in internal RAM
	.equ flag,00h	;addressable bit 00 used as a flag
	.org 0000h	
getkey:	mov pl,#0ffh	;set ports 1 and 3 as inputs
	mov p3,#0ffh	
scan:	acall keydown	;keydown looks for any key(s) down
	jz scan	;if A = 0 then no key(s) down; loop
	acall convert	;convert returns flag set if not ;valid
	jbc flag,scan	;or A set to 00 to 09 for keys 0–9
	mov newkey,a	;store key and wait for debounce ;time
	mov a,#bounce	;then check to see if <u>same</u> key
	acall softime	;wait 20 ms
	acall keydown	;see if a key is <u>still</u> down
	jz scan	;if not down then must have been ;noise
	acall convert	;see if key is still valid and ;matches
	jbc flag,scan	;the original key found
	cjne a,newkey,scan	;check for equal
	acall vendit	;call vending machine subroutine
wait:	acall keydown	;now wait for all keys to go up
	jnz wait	;wait until A = 00: keys all up
	mov a,#next	;wait 50d ms and see if all ;still up
	acall softime	
	acall keydown	;continue until keys are up
	jnz wait	;loop until keys up for 50d ms
	sjmp scan	;get next key

```
;
;"keydown" gets the contents of Pl and P3 pins, which are connected
;to the keys, and checks for any zero bits; no check is made to see
;if more than one bit is low
;
```

keydown:	mov r0,pl	;get state of Pl keys to R0
	mov a,p3	;get state of P3 keys to A
	orl a,#0f3h	;make bits 0,1,and 4–7 a one
	anl a,r0	;check for any one or more keys ;down
	cpl a	;A = FFh if all keys up, now 00
	ret	;if A not 00 then at least one key ;down

Continued

ADDRESS	MNEMONIC	COMMENT

Continued

```
;"convert" checks for more than one key down; if more than one key
;is down then addressable bit "flag" is set; if only one key is
;down then the one-of-ten bit pattern is converted to an
;equivalent 0-9 number in the A register and "flag" is reset
;valid patterns (a single 0 out of ten bits) are found by CJNE
;operations; A is counted up for each test to match the key number
;
convert:    clr flag              ;assume that key hit is valid
            clr a                 ;A contains first possible key (00)
            mov r1,p1             ;get P1 key pattern in R1
            mov r3,p3             ;get P3 key pattern in R3
            orl 03h,#0f3h         ;make r3 bits 0,1 and 4-7 a one
            cjne r1,#0feh,one     ;search R1 for a legal pattern
            sjmp check3           ;check R3 for no key down
one:        inc a                 ;A contains next key possible (01)
            cjne r1,#0fdh,two     ;continue this for all valid
                                  ;patterns

            sjmp check3
two:        inc a                 ;A = 02
            cjne r1,#0fbh,three
            sjmp check3
three:      inc a                 ;A = 03
            cjne r1,#0f7h,four
            sjmp check3
four:       inc a                 ;A = 04
            cjne r1,#0efh,five
            sjmp check3
five:       inc a                 ;A = 05
            cjne r1,#0dfh,six
            sjmp check3
six:        inc a                 ;A = 06
            cjne r1,#0bfh,seven
            sjmp check3
seven:      inc a                 ;A = 07
            cjne r1,#7fh,eight
            sjmp check3
eight:      inc a                 ;A = 08
            cjne r3,#0fdh,nine    ;now look for a key in R3
            jnb p3.3 bad          ;check that key 9 is up
            sjmp good
nine:       inc a                 ;A = 09
            cjne r3,#0f7h,bad     ;redundant check
good:       ret
check3:     jnb p3.3,bad          ;if R1 has a low then P3 must be
                                  ;  high

            jnb p3.4,bad
            sjmp good
```

Continued

ADDRESS	MNEMONIC	COMMENT
bad:	setb flag	;signal an invalid key pattern
	ret	
softime:	ret	;simulate ''softime'' subroutine
vendit:	ret	;simulate ''vendit'' subroutine
	.end	

COMMENT

The "convert" subroutine is looking for a single low bit. The CJNE patterns all have one bit low and the rest high.

Multiple keys are rejected by "convert." Held keys are ignored as the program waits for a 50d millisecond "all keys up" period before admitting the next key. The program loops so quickly that it is humanly impossible to hit a key so that it can be missed.

The main program is predominantly a series of calls to subroutines which can each be written by different programmers. Agreement on what data is passed to and received from the subroutines is essential for success, as well as a clear understanding of what 8051 registers and memory locations are used.

Interrupt-Driven Programs for Small Keyboards

If the application is so time sensitive that the delays associated with debouncing and awaiting an "all-up" cannot be tolerated, then some form of interrupt must be used so that the main program can run unhindered.

A compromise may be made by polling the keyboard as the main program loops, but all time delays are done using timers so that the main program does not wait for a software delay. The "Getkey" program can be modified to use a timer to generate the delays associated with the key down debounce time and the "all-up" delay. The challenge associated with this approach is to have the program remember which delay is being timed out. Remembering which delay is in progress can be handled using a flag bit, or one timer can be used to generate the key-down debounce delay, and another timer to generate the key-up delay. The flag approach is examined in the example given in this section.

The important feature of the program is that the main program will check a flag to see whether there is any keyboard activity. If the flag is set, then the program finds the key stored in a RAM location and resets the flag. The getting of the key is "transparent" to the main program; it is done in the interrupt program. The keyboard is still polled by the main program, but the interrupt program gets the key after that. The program named "Hardtime" from Chapter 7 is used for the time delay. The keyboard user may notice some sluggishness in response if the main program takes so long to loop that the keyboard initiation sequence is not done every quarter-second or so.

Inkey

The program "Inkey" uses hardware timer T1 to generate all time delays. The keyboard sequence is initiated when a key is found to be down; otherwise, the program continues and checks for a key down in the next loop. A key down initiates a debounce time delay in timer T1 and sets a timer flag to notify the interrupt program of the use of the timer. The interrupt program checks that a key is still down and is valid. Valid keys are stored, and a flag is set that may be tested by the main program. The interrupt program then begins the key-up delay and sets the timer flag to signify this condition. After each key-up delay, the interrupt program checks for all keys up. The time delay is reinitialized until all keys are up and the timer interrupts are halted.

ADDRESS	MNEMONIC	COMMENT
	.equ newkey,70h	;store any new key in RAM
	.equ flag,00h	;addressable bit 00 used as a flag
	.equ newflg,01h	;when newflg = 1 then there is ;a key
	.equ timflg,02h	;timflg = 0 for debounce, 1 for ;delay
	.equ bounce,14h	;set debounce delay to 20d ms
	.equ next,32h	;set interval between keys to ;50d ms
	.equ savetime,0010h	;external RAM address for delay
	.org 0000h	
inkey:	.sjmp over	;jump over interrupt locations

```
;
;when T1 times out it vectors here and jumps to ''hardtime'' for the
;desired delay. When the delay is up then the key program is called,
;
```

	.org 001bh	;interrupt location for TF1
	ljmp hardtime	;jump to time delay subroutine
userdly:	acall usertime	;call usertime if delay done
	reti	;return to program when usertime ;done

```
;
;the main program begins here; the keyboard is scanned unless there
;is a new key to be processed, or T1 is counting, signifying that
;a key read is in progress
;
```

over:	mov p1,#0ffh	;set ports 1 and 3 as inputs
	mov p3,#0ffh	
	clr newflg	;initialize all flags
	clr flag	
	clr timflg	
begin:	jbc newflg,key	;check if a key is waiting and ;get it
	jb tcon.6,mainprog	;if T1 is running then wait
	acall keydown	;keydown looks for any key(s) down
	jz mainprog	;if A = 0 then no key(s) down; ;go on
	acall convert	;check for a valid key
	jz mainprog	;go on with main program if not ;valid
	mov newkey,a	;store key and start debounce timer
	clr timflg	;signal interrupt program T1 ;running
	mov dptr,#savetime	;point to delay address
	mov a,#bounce	;set 20 ms delay
	movx @dptr,a	
	inc dptr	;point to next byte
	mov a,#00h	
	movx @dptr,a	;desired delay now stored

Continued

ADDRESS	MNEMONIC	COMMENT

```
                 orl ie,#88h            ;enable interrupts and Tl interrupt
                 acall startime         ;start time delay go to mainprog
                 sjmp mainprog
key:             mov a,newkey           ;get key and use in main program
mainprog:        sjmp begin             ;simulate main program and
                                        ;loop back
;
; ***************************** CONVERT ****************************
;"convert" checks for more than one key down; if more than one key
;is down then addressable bit "flag" is set; if only one key is
;down then the one-of-ten bit pattern is converted to an
;equivalent 0-9 number in the A register and "flag" is reset
;valid patterns (a single 0 out of ten bits) are found by CJNE
;operations; A is counted up for each test to match the key number
convert:         clr flag               ;assume that key hit is valid
                 clr a                  ;A contains first possible key (00)
                 mov rl,pl              ;get Pl key pattern in Rl
                 mov r3,p3              ;get P3 key pattern in R3
                 orl 03h,#0f3h          ;make R3 bits 0,1 and 4-7 a one
                 cjne rl,#0feh,one      ;search Rl for a legal pattern
                 sjmp check3            ;check R3 for no key down
one:             inc a                  ;A contains next key possible (01)
                 cjne rl,#0fdh,two      ;continue this for all valid
                                        ;patterns
                 sjmp check3
two:             inc a                  ;A = 02
                 cjne rl,#0fbh,three
                 sjmp check3
three:           inc a                  ;A = 03
                 cjne rl,#0f7h,four
                 sjmp check3
four:            inc a                  ;A = 04
                 cjne rl,#0efh,five
                 sjmp check3
five:            inc a                  ;A = 05
                 cjne rl,#0dfh,six
                 sjmp check3
six:             inc a                  ;A = 06
                 cjne rl,#0bfh,seven
                 sjmp check3
seven:           inc a                  ;A = 07
                 cjne rl,#7fh,eight
                 sjmp check3
eight:           inc a                  ;A = 08
                 cjne r3,#0fdh,nine     ;now look for a key in R3
                 jnb p3.3 bad           ;check key 9 is up
                 sjmp good
nine:            inc a                  ;A = 09
                 cjne r3,#0f7h,bad      ;redundant check
```

Continued

ADDRESS	MNEMONIC	COMMENT

Continued

```
good:       ret
check3:     jnb p3.3,bad            ;if R1 has a low then P3 must
                                    ;be high
            jnb p3.4,bad
            sjmp good
bad:        setb flag               ;signal an invalid key pattern
            ret
;
; *************************** HARDTIME ***************************
;"hardtime" will count the interrupts generated by T1 until the
;number placed in RAM location "savetime" is zero
;
hardtime:   push acc                ;save registers to be used
            push dph
            push dpl
            mov dptr,#savetime      ;get pointer to time delay number
            movx a,@dptr            ;count delay number down to 0000
            dec a                   ;low byte first
            cjne a,#00h,aff         ;check for 0000
            movx @dptr,a            ;save low byte = 00
            inc dptr                ;get high byte and look for 00
            movx a,@dptr
            jz done                 ;done if low byte = high byte = 00
            sjmp sava               ;not 0000, reset T1 and delay again
aff:        cjne a,#0ffh,sava       ;if low byte is FF then dec high
            movx @dptr,a            ;save low byte = FF
            inc dptr                ;point to high byte
            movx a,@dptr            ;count high byte down
            dec a
            sjmp sava               ;save the high byte
done:       pop dpl                 ;delay is finished
            pop dph                 ;restore all registers used
            pop acc
            ljmp userdly            ;continue at user delay program
sava:       movx @dptr,a            ;delay is not up, save the byte
            pop dpl                 ;restore saved registers
            pop dph
            pop acc
            acall startime          ;start T1 for next 1 ms delay
            reti                    ;return to place in user program
;
; *************************** KEYDOWN ***************************
;"keydown" gets the contents of P1 and P3 pins that are connected
;to the keys and checks for any zero bits; no check is made to see
;if more than one bit is low
;
keydown:    mov r0,p1               ;get state of P1 keys to R0
```

Continued

ADDRESS	MNEMONIC	COMMENT

```
                mov a,p3              ;get state of P3 keys to A
                orl a,#0f3h           ;make bits 0,1,and 4-7 a one
                anl a,r0              ;check for any one or more
                                      ;keys down
                cpl a                 ;A = FFh if all keys up, now 00
                ret                   ;if A not 00 then one or more down
;
; *************************** STARTIME ***************************
;"startime" initializes timer 1 and enables timing to begin
;
startime:       mov thl, #0fah        ;set T1 for a 1 ms delay
                mov tll, #0cbh        ;(see "Timer" example in
                                      ;Chapter 7)
                anl tmod,#0fh         ;clear T1 part of TMOD
                orl tmod,#40h         ;set T1 to timer mode 1
                orl tcon,#40h         ;start timer 1
                ret                   ;return to calling program
;
; *************************** STOPTIME ***************************
;"stoptime" disables T1
;
stoptime:       anl tcon,#0bfh        ;stop timer T1
                ret
;
; *************************** USERTIME ***************************
;"usertime,"the user program called from the interrupt program after
;hardtime has timed out, will process the key and set the 50d ms
;delay if the key was valid
;
usertime:       acall stoptime        ;stop timer and determine T1 use
                jb timflg,keyup       ;if a delay then see if keys up
                acall keydown         ;see if a key is still down
                jz goback             ;if not down then must be noise
                acall convert         ;see if key is valid and matches
                jbc flag,goback       ;the original key found
                cjne a,newkey,goback
                setb newflg           ;set new key flag for main program
delay:          mov dptr,#savetime    ;point to delay address
                mov a,#next           ;set 50d ms delay
                movx @dptr,a
                inc dptr              ;point to next byte
                mov a,#00h
                movx @dptr,a          ;desired delay now stored
                orl ie,#88h           ;enable interrupts and T1 interrupt
                acall startime        ;start time delay
                setb timflg           ;set flag for delay condition
goback:         ret
```

Continued

ADDRESS	MNEMONIC	COMMENT
Continued		
keyup:	acall keydown	;see if keys are up after delay
	jnz delay	;if not then delay again
	sjmp goback	;return with T1 stopped
	.end	

────▷── COMMENT ──────────────────────────────

This program is large enough to require additional attempts to make it legible. All of the sub-routines are arranged in alphabetical order.

Codekey

The completely interrupt-driven small keyboard example given in this section requires no program action until a key has been pressed. Hardware must be added to attain a completely interrupt-driven event. The circuit of Figure 8.4 is used.

FIGURE 8.4 Keyboard Configuration Used for "Codekey" Program

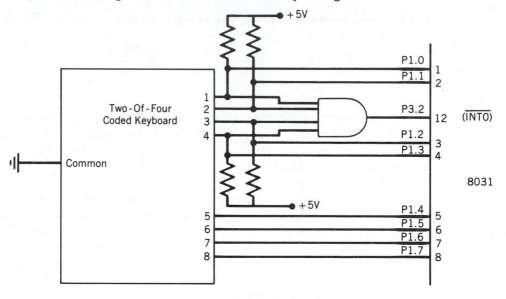

Keyboard Code		
Key	Pins	Low
0	1	5
1	2	5
2	3	5
3	4	5
4	1	6
5	2	6
6	3	6
7	4	6
8	1	7
9	2	7

The keyboard is a two-of-eight type which codes the ten keys as follows:

KEY	CODE(HEX)
0	EE
1	ED
2	EB
3	E7
4	DE
5	DD
6	DB
7	D7
8	BE
9	BD

An inspection of the code reveals that each nibble has only one bit that is low for each key and that two of the eight bits are uniquely low for each key. If more than one key is pressed, then three or more bits go low, signaling an invalid condition. This popular scheme allows for up to 16 keys to be coded in this manner. Unlike the lead-per-key arrangement, only four of the lines must be active-low ORed to generate an interrupt.

The hardware serves to detect when any number of keys are hit by using an AND gate to detect when any nibble bit goes low. The high-to-low transition then serves to interrupt the microcontroller on port 3.2 ($\overline{INT0}$). The interrupt program reads the keys connected to port 1 and uses timer T0 to generate the debounce time and T1 for the keys-up delay. The total delay possible at 16 megahertz for the timers is 49.15 milliseconds, which covers the delay times used in the previous examples.

The program "Codekey" which is interrupt driven by a high-to-low transition on $\overline{INT0}$. Timers T0 and T1 generate the debounce and delay times in an interrupt mode. The $\overline{INT0}$ interrupt input is disabled until all keys have been seen up for the T1 delay. A lookup table is used to verify that only one key is pressed.

ADDRESS	MNEMONIC	COMMENT
	.equ newkey,70h	;store a new key in RAM
	.equ base,400h	;base of lookup table
	.equ newflg,00h	;addressable bit 00 for new key flag
	.org 0000h	
codekey:	sjmp over	;jump over interrupt locations
	.org 0003h	;this is the $\overline{INT0}$ interrupt vector
	sjmp keyint	
	.org 000bh	;timer T0 interrupt vector
	sjmp tim0	
	.org 001bh	;timer T1 interrupt vector
	sjmp tim1	
keyint:	mov tl0,#0d4h	;set T0 for 20 ms delay
	mov th0,#97h	;count from 97D4h to 0000
	setb tcon.4	;start timer T0
	clr ie.0	;disable $\overline{INT0}$ interrupt
	reti	;enable interrupt structure and ;return
tim0:	push acc	;save registers used
	push dpl	

Continued

ADDRESS	MNEMONIC	COMMENT
Continued		
	push dph	
	clr tcon.4	;stop T0
	mov a,pl	;get key pattern
	mov dptr,#base	;set DPTR to point to lookup table
	movc a,@a+dptr	;not valid = FFh
	cjne a,#0ffh,good	
	pop dph	
	pop dpl	
	pop acc	
	setb ie.0	;enable $\overline{INT0}$ interrupt
	reti	;enable interrupt structure and
		;return
good:	mov newkey,a	;store the newkey
	setb newflg	;signal main program; new key present
	anl tll,#00h	;set Tl for maximum delay (49.1 ms)
	anl thl,#00h	
	setb tcon.6	;start timer Tl
	pop dph	;restore retgisters
	pop dpl	
	pop acc	
	reti	;enable interrupt structure and
		;return
Timl:	push acc	;save A
	clr tcon.6	;stop Tl
	mov a,pl	;see if keys up yet
	cjne a,#0ffh,wait	;all inputs will be high if all up
	setb ie.0	;enable $\overline{INT0}$ for next key
	pop acc	
	reti	
wait:	anl tll,#00h	;restart Tl and delay again
	anl thl,#00h	
	setb tcon.6	;start Tl
	pop acc	
	reti	;return with interrupt enabled
over:	mov tcon,#01h	;set $\overline{INT0}$ for falling edge interrupt
	mov ie,#8bh	;enable $\overline{INT0}$, T0, and Tl interrupts
	mov tmod,#11h	;choose timer operation; mode 1
simulate:	jbc newflg,key	;see if there is a new key and get it
	sjmp simulate	;simulate rest of program
key:	mov a,newkey	;get key and simulate rest of program
	sjmp simulate	
	.org 04bdh	;place lookup table here, keys 9
		;and 8
	.db 09h	
	.db 08h	

Continued

ADDRESS	MNEMONIC	COMMENT
	.org 04d7h	;key 7
	.db 07h	
	.org 04dbh	;key 6
	.db 06h	
	.org 04ddh	;keys 5 and 4
	.db 05h	
	.db 04h	
	.org 04e7h	;key 3
	.db 03h	
	.org 04ebh	;key 2
	.db 02h	
	.org 04edh	;keys 1 and 0
	.db 01h	
	.db 00h	
	.end	

 COMMENT

The lookup table will work only if every bit from 0400h to 04FFh that is not a .db assignment is FFh. Most EPROMS will be FFh when erased, and the assembler will not program unspecified locations. The table will have to be assembled so that an FFh is at every non-key location if this is not true.

Key bounce down is eliminated by the T0 delay, and key bounce up, by the T1 delay. More than two keys down is detected by the self-coding nature of the keyboard. A held key does not interrupt the edge-triggered $\overline{INT0}$ input.

Program for a Large Matrix Keyboard

A 64-key keyboard, arranged as an 8-row by 8-column matrix will be interfaced to the 8051 microcontroller, as shown in Figure 8.5. Port 1 will be used to bring each row low, one row at a time, using an 8-bit latch that is strobed by port 3.2. P1 will then read the 8-bit column pattern by enabling the tri-state buffer from port 3.3. A pressed key will have a unique row-column pattern of one row low, one column low. Multiple key presses are rejected by either an invalid pattern or a failure to match for three complete cycles. Each row is scanned at an interval of 1 millisecond, or an 8 millisecond cycle for the entire keyboard. A valid key must be seen to be the same key for 3 cycles (24 milliseconds). There must then be three cycles with no key down before a new key will be accepted. The 1 millisecond delay between scans is generated by timer T0 in an interrupt mode.

Bigkey

The "Bigkey" program scans an 8 × 8 keyboard matrix using T0 to generate a periodic 1 ms delay in an interrupt mode. Each row is scanned via an external latch driven by port 1 and strobed by port 3.2. Columns are read via a tri-state buffer under control of port 3.3. Keys found to be valid are passed to the main program by setting the flag "newflg" and placing the key identifiers in locations "newrow" and "newcol." The main program resets "newflg" when the new key is fetched. R4 is used as a cycle counter for successful matches and up time cycles. R5 is used to hold the row scan pattern: only one bit low.

FIGURE 8.5 Circuit for "Bigkey" Program

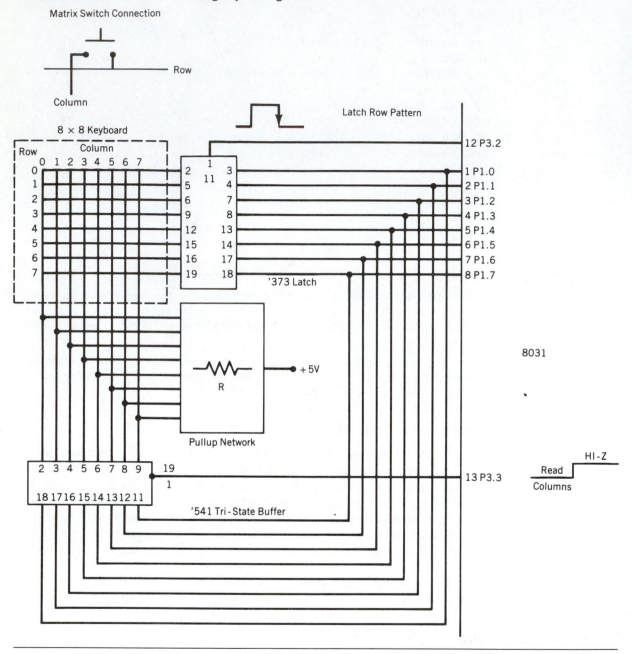

ADDRESS	MNEMONIC	COMMENT
	.equ newrow,70h	;store any valid key row address
	.equ newcol,71h	;store any valid key column address
	.equ newflg,00h	;use addressable bit as a new
		;key flag

Continued

ADDRESS	MNEMONIC	COMMENT
	.equ upflg,01h	;upflag signals start of key up ;delay
	.org 0000h	
bigkey:	sjmp over	;jump over T0 interrupt to main ;program
;		
;The interrupt program begins here; T0 is reloaded to permit		
;the next interrupt in 1 ms		
;		
	.org 000bh	;vector location for T0 overflow ;flag
	mov tl0,#0cbh	;reload T0 for next interrupt
	mov th0,#0fah	
	push acc	;save A and the flags
	push psw	
	mov p1,r5	;get row scan pattern to port 1
	setb p3.2	;generate a strobe to the latch
	clr p3.2	
	mov p1,#0ffh	;set P1 as an input port
	clr p3.3	;read buffer and see if any ;key down
	mov a,p1	;get column pattern
	setb p3.3	;disable buffer
	jb upflg,upyet	;if upflg = 1 then wait for ;keys up
	setb c	;set C to 1 and rotate A to find ;a low
	mov r3,#08h	;8 rotates will restore A to ;original
look:	rrc a	;see if only one zero in A (valid)
	jnc test	;if C = 0 then see if A = FFh
	djnz r3,look	;go until C = 0 or rotate finished
	mov a,r5	;check for a key down previous scan
	cjne a,newrow,goback	
	mov newrow,#00h	;if so then not repeated; zero ;newrow
	mov r4,#00h	;if so then zero R4 and scan again
	sjmp goback	;return to main program
test:	cjne a,#0ffh,bad	;if A not all ones then invalid key
here:	rrc a	;good pattern; restore A
	djnz r3,here	
	cjne r4,#00h,match	;R4 counts pattern matches
newone:	mov newcol,a	;first time seen; see if it recurs
	mov newrow,r5	
	inc r4	;R4 contains key detected count
	sjmp goback	
match:	push acc	;save A and check R5 for a new row
	mov a,r5	

Continued

ADDRESS	MNEMONIC	COMMENT
Continued		
	cjne a,newrow,unk	;if no match then this is a new key
	pop acc	;restore A and check for a new
		;column
	cjne a,newcol,unkn	;if no match then this is a new key
	inc r4	;match: see if 24 ms have expired
	cjne r4,#04h,goback	;keep if seen for at least 3 cycles
good:	mov newrow,r5	;save new key row and column
	mov newcol,a	
	setb newflg	
	setb upflg	;set up flag for 3 cycles up
	mov r4,#00h	;reset R4 to count key up cycles
	sjmp goback	
unk:	pop acc	;restore new column pattern to A
unkn:	mov r4,#00h	;reset r4 to reflect a new key
	sjmp newone	;look for matches on next cycles
bad:	mov r4,#00h	;reset match counter
	sjmp goback	
upyet:	cjne a,#0ffh,notup	;look for A = FFh
	inc r4	;R4 now counts 3 cycle of up time
	cjne r4,#18h,goback	;look for 24d scans (3 cycles)
	clr upflg	;up time done, look for next key
notup:	mov r4,#00h	;reset R4
goback:	mov a,r5	;rotate R5 low bit to next row
	rl a	
	mov r5,a	
	pop psw	;restore PSW and A
	pop acc	
	reti	

```
;
;the interrupt program finishes here and the main program begins;
;the main program would normally get the new key row and column
;patterns and convert these to a single byte number
;
```

over:	mov r5,#0feh	;initialize R5 for bottom row low
	mov tmod,#01h	;set T0 to mode 1
	mov tl0,#0cbh	;set T0 for a 1 ms delay
	mov th0,#0fah	;count 1333d @ .75 μs/count
	mov ie,#82h	;enable the T0 interrupt
	setb tcon.4	;start timer
	mov r4,#00h	;reset R4 for no valid key
	clr upflg	;reset key up flag
	clr newflg	;reset new key flag
main:	jbc newflg,simulate	;get key row and column addresses
	sjmp main	;simulate main program
simulate:	nop	;main program would get addresses
		;here
	sjmp main	
	.end	

COMMENT

Once begun by the main program, T0 continues to time out and generate the row scan pattern in the interrupt program. To the main program, the keys appear in some unknown way; the interrupt program is said to run in the "background."

There is considerable adjustment (tweak) in this program to accommodate keys with various bounce characteristics. The debounce time can be altered in a gross sense by changing the number of cycles (R4) for acceptance and in a fine way by changing the basic row scan time (T0).

This same program can be used to monitor any multipoint array of binary data points. The array can be expanded easily to a 16 × 16 matrix by adding one more latch and tristate buffer and using two more port 3 pins to generate the latch and enable strobes.

Note that only A can compare against memory contents in a CJNE instruction.

Displays

If keyboards are the predominant means of interface to human input, then visible displays are the universal means of human output. Displays may be grouped into three broad categories:

1. Single light(s)
2. Single character(s)
3. Intelligent alphanumeric

Single light displays include incandescent and, more likely, LED indicators that are treated as single binary points to be switched off or on by the program. *Single character* displays include numeric and alphanumeric arrays. These may be as simple as a seven-segment numeric display up to intelligent dot matrix displays that accept an 8-bit ASCII character and convert the ASCII code to the corresponding alphanumeric pattern. *Intelligent alphanumeric* displays are equipped with a built-in microcontroller that has been optimized for the application. Inexpensive displays are represented by multicharacter LCD windows, which are becoming increasingly popular in hand-held wands, factory floor terminals, and automotive dashboards. The high-cost end is represented by CRT ASCII terminals of the type commonly used to interface to a multi-user computer.

The individual light and intelligent single-character displays are easy to use. A port presents a bit or a character then strobes the device. The intelligent ASCII terminals are normally serial devices, which are the subject of Chapter 9.

The two examples in this section—seven-segment and intelligent LCD displays—require programs of some length.

Seven-Segment Numeric Display

Seven-segment displays commonly contain LED segments arranged as an "8," with one common lead (anode or cathode) and seven individual leads for each segment. Figure 8.6 shows the pattern and an equivalent circuit representation of our example, a common cathode display. If more than one display is to be used, then they can be time multiplexed; the human eye can not detect the blinking if each display is relit every 10 milliseconds or so. The 10 milliseconds is divided by the number of displays used to find the interval between updating each display.

The example examined here uses four seven-segment displays; the segment information is output on port 1 and the cathode selection is done on ports 3.2 to 3.5, as shown in

FIGURE 8.6 Seven-Segment LED Display and Circuit

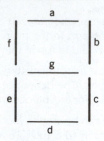

Segment Pattern

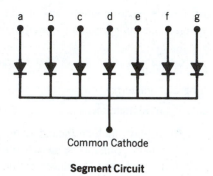

Common Cathode

Segment Circuit

Figure 8.7. A segment will be lit *only if* the segment line is brought high *and* the common cathode is brought low.

Transistors must be used to handle the currents required by the LEDs, typically 10 milliamperes for each segment and 70 milliamperes for each cathode. These are average current values; the peak currents will be four times as high for the 2.5 milliseconds each display is illuminated.

The program is interrupt driven by T0 in a manner similar to that used in the program "Bigkey." The interrupt program goes to one of four two-byte character locations and finds the cathode segment pattern to be latched to port 1 and the anode pattern to be latched to port 3. The main program uses a lookup table to convert from a hex number to the segment pattern for that number. In this way, the interrupt program automatically displays whatever number the main program has placed in the character locations. The main program loads the character locations and is not concerned with how they are displayed.

Svnseg

The program "svnseg" displays characters found in locations "ch1" to "ch4" on four common-cathode seven-segment displays. Port 1 holds the segment pattern from the low byte of chx; port 3 holds the cathode pattern from the high byte of chx. T0 generates a 2.5 ms delay interval between characters in an interrupt mode. The main program uses a lookup table to convert from hex to a corresponding pattern. R0 of bank one is dedicated as a pointer to the displayed character.

FIGURE 8.7   Seven-Segment Display Circuit Used for "Svnseg" Program

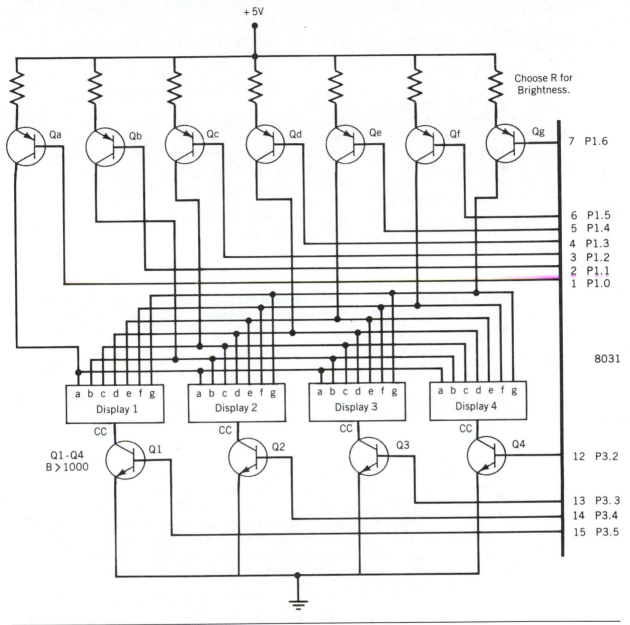

ADDRESS	MNEMONIC	COMMENT
	.equ ch1,50h	;assign RAM character locations
	.equ ch2,52h	;two bytes per character
	.equ ch3,54h	
	.equ ch4,56h	
	.org 0000h	;jump over T0 interrupt location
svnseg:	mov sp,#0fh	;get the stack above bank one
	sjmp over	

Continued

ADDRESS	**MNEMONIC**	**COMMENT**

Continued

```
;
;begin the interrupt-driven program at the T0 interrupt location
;
            .org 000bh
            mov tl0,#0fbh        ;reload T0 for next interrupt
            mov th0,#0f2h
            setb psw.3           ;select bank one
            mov pl,@r0           ;place segment pattern on port 1
            inc r0               ;point to accompanying cathode pattern
            mov p3,@r0           ;place cathode patten on port 3
            inc r0               ;check for fourth character
            cjne r0,#58h,nxt
            mov r0,#chl          ;if ch4 just displayed go to chl
nxt:        clr psw.3            ;return to register bank 0
            reti                 ;return to main program
;the main program loads sample characters and starts the T0
;interrupt.
            mov a,#00h           ;use an example sequence of 0, 1, 2, 3
            acall convert        ;convert to segment pattern and store
            mov chl,a
            mov a,#01h
            acall convert
            mov ch2,a
            mov a,#02h
            acall convert
            mov ch3,a
            mov a,#03h
            acall convert
            mov ch4,a            ;last segment pattern stored
            setb psw.3           ;select register bank one
            mov r0,#chl          ;set R0 to point to chl RAM location
            inc r0               ;now load anode pattern for chl
            mov @r0,#20h         ;set anode for character 1 only high
            inc r0               ;point to next character and continue
            inc r0               ;load ch2 pattern
            mov @r0,#10h
            inc r0
            inc r0               ;load ch3 pattern
            mov @r0,#08h
            inc r0
            inc r0               ;load ch4 pattern
            mov @r0,#04h
            mov r0,#chl          ;point to RAM address for chl
            mov tl0,#0fbh        ;load T0 for first interrupt
            mov th0,#0f2h
```

Continued

ADDRESS	MNEMONIC	COMMENT

```
                mov tmod,#01h       ;set T0 to mode 1
                mov ie,#82h         ;enable T0 interrupt
                setb tcon.4         ;start timer
                clr psw.3           ;return to register bank 0
here:           sjmp here           ;loop and simulate rest of program
;
;convert uses the PC to point to the base of the 16-byte table
;
convert:        inc a               ;compensate for RET byte
                mov a,@pc+a         ;get byte
                ret                 ;return with segment pattern in A
                .db c0h             ;0
                .db f9h             ;1
                .db a4h             ;2
                .db b0h             ;3
                .db 99h             ;4
                .db 92h             ;5
                .db 82h             ;6
                .db f8h             ;7
                .db f0h             ;8
                .db 98h             ;9
                .db 88h             ;A
                .db 83h             ;b
                .db c6h             ;C
                .db b1h             ;d
                .db 86h             ;E
                .db 8eh             ;F
                .end
```

▷ COMMENT ───

Using bank 1 as a dedicated bank for the interrupt routine cuts down on the need for pushes and pops. Bank 1 may be selected quickly, giving access to the eight registers while saving the bank 0 registers. Note that the stack, at reset, points to R0 of bank 1, so that it must be relocated.

The intensity of the display may also be varied by blanking the displays completely for some interval using the program.

Intelligent LCD Display

In this section, we examine an intelligent LCD display of two lines, 20 characters per line, that is interfaced to the 8051. The protocol (handshaking) for the display is shown in Figure 8.8, and the interface to the 8051 in Figure 8.9.

The display contains two internal byte-wide registers, one for commands (RS = 0) and the second for characters to be displayed (RS = 1). It also contains a user-programmed RAM area (the character RAM) that can be programmed to generate any desired character that can be formed using a dot matrix. To distinguish between these two data areas, the hex command byte 80 will be used to signify that the display RAM address 00h is chosen.

FIGURE 8.8 Intelligent LCD Display

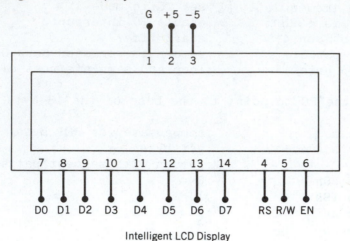

Intelligent LCD Display

BIT	RS	R/W	D7	D6	D5	D4	D3	D2	D1	D0	Function
	0	0	0	0	0	0	0	0	0	1	Clear LCD and memory, home cursor
	0	0	0	0	0	0	0	0	1	0	Clear and home cursor only
	0	0	0	0	0	0	0	1	I/O	S	Screen action as display character written
											S = 1/0: Shift screen/cursor
											I/O = 1/0: Cursor R/L, screen L/R
	0	0	0	0	0	0	1	D	C	B	D = 1/0: Screen on/off
											C = 1/0: Cursor on/off
											B = 1/0: Cursor Blink/Noblink
	0	0	0	0	0	1	S/C	R/L	0	0	S/C = 1/0: Screen/Cursor
											R/L = 1/0: Shift one space R/L
	0	0	0	0	1	DL	N	F	0	0	DL = 1/0: 8/4 Bits per character
											N = 1/0; 2/1 Rows of characters
											F = 1/0; 5X10/5X7 Dots/Character
	0	0	0	1	Character address						Write to character RAM Address after this
	0	0	1	Display data address							Write to display RAM Address after this
	0	1	BF	Current address							BF = 1/0: Busy/Notbusy
	1	0	Character byte								Write byte to last RAM chosen
	1	1	Character byte								Read byte from last RAM chosen

Port 1 is used to furnish the command or data byte, and ports 3.2 to 3.4 furnish register select and read/write levels.

The display takes varying amounts of time to accomplish the functions listed in Figure 8.8. LCD bit 7 is monitored for a logic high (busy) to ensure the display is not overwritten. A slightly more complicated LCD display (4 lines × 40 characters) is currently being used in medical diagnostic systems to run a very similar program.

Lcdisp

The program "lcdisp" sends the message "hello" to an intelligent LCD display shown in Figure 8.8. Port 1 supplies the data byte. Port 3.2 selects the command (0) or data (1) registers. Port 3.3 enables a read (0) or write (1) level, and port 3.4 generates an active low-enable strobe.

FIGURE 8.9 Intelligent LCD Circuit for "Lcdisp" Program

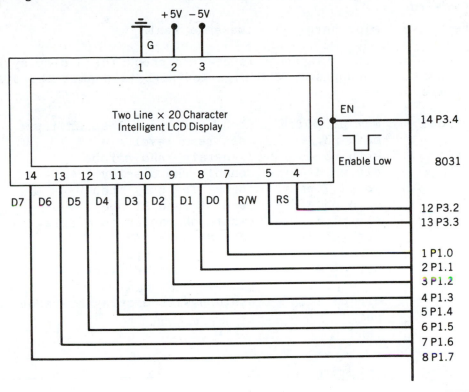

ADDRESS	MNEMONIC	COMMENT
	.org 0000h	
lcdisp:	clr p3.2	;select the command register
	clr p3.3	;select write level
	mov a,#3fh	;command 8 bits/char., 2 rows, 5 x 10
	acall strobe	;strobe command to display
	mov a,#0eh	;command screen and cursor on, no blink
	acall strobe	
	mov a,#06h	;command cursor right as data displayed
	acall strobe	
	mov a,#01h	;clear all and home cursor
	acall strobe	
	setb p3.2	;select display data RAM register
	mov a,#'h'	;say "hello"
	acall strobe	
	mov a,#'e'	
	acall strobe	
	mov a,#'l'	
	acall strobe	
	acall strobe	
	mov a,#'o'	
	acall strobe	

Continued

ADDRESS	MNEMONIC	COMMENT

Continued

```
here:       sjmp here          ;message sent
;
;the subroutine ''strobe'' is used to check for a display busy
;condition, and pulse P3.3 high-low-high to enable the display
;write or read
;
strobe:     mov p1,#0ffh       ;configure port 1 as an input
            setb p3.3          ;set read level
wait:       setb p3.4          ;generate read strobe
            clr p3.4           ;enable the display
            jb p1.7,wait       ;check for busy when BF = 1
            setb p3.4          ;end of read strobe
            clr p3.3           ;write character to display
            setb p3.2          ;choose data RAM
            mov p1,a           ;character to port 1
            clr p3.4           ;generate write strobe
            setb p3.4
            clr p3.2           ;return with display as before call
            ret
            .end
```

COMMENT

If long character strings are to be displayed, then a subroutine could be written that receives the beginning address of the string. The subroutine then displays the characters until a unique "end-of-string" character is found.

Pulse Measurement

Sensors used for industrial and commercial control applications frequently produce pulses that contain information about the quantity sensed. Varying the sensor output frequency, using a constant duty cycle but variable frequency pulses to indicate changes in the measured variable, is most common. Varying the duration of the pulse width, resulting in constant frequency but variable duty cycle, is also used. In this section, we examine programs that deal with both techniques.

Measuring Frequency

Timers T0 and T1 can be used to measure external frequencies by configuring one timer as a counter and using the second timer to generate a timing interval over which the first can count. The frequency of the counted pulse train is then

$$\text{Unknown frequency} = \text{Counter/timer}$$

For example, if the counter counts 200 pulses over an interval of .1 second generated by the timer, the frequency is

$$\text{UF} = 200/.1 = 2000 \text{ Hz}$$

Certain fundamental limitations govern the range of frequencies that can be measured. An input pulse must make a 1-to-0 transition lasting two machine cycles, or f/24, to be counted. This restriction on pulse deviation yields a frequency of 667 kilohertz using our 16 megahertz crystal (assuming a square wave input).

The lowest frequency that can be counted is limited by the duration of the time interval generated, which can be exceedingly long using all the RAM to count timer rollovers (49.15 milliseconds $\times 2^{\wedge}32768$). There is no practical limitation on the lowest frequency that can be counted.

Happily, most frequency variable sensors generate signals that fall inside of 0 to 667 kilohertz. Usually the signals have a range of 1,000 to 10,000 hertz.

Our example will use a sensor that measures dc voltage from 0 to 5 volts. At 0 V the sensor output is 1,000 hertz, and at full scale, or 5 volts, the sensor output is 6,000 hertz. The correspondence is 1 volt per 1,000 hertz, and we wish to be able to measure the voltage to the nearest .01 V, or 10 hertz of resolution (assuming the sensor is this accurate). A timing interval of 1 second generates a frequency count accurate to the nearest 1 hertz, so an interval of .1 s yields a count accurate to the nearest 10 hertz.

Another way to arrive at the desired timing interval, T, is to note that the desired accuracy is

$$\frac{.01 \text{ V}}{5 \text{ V}} = .002 = \frac{1}{512} = \frac{1}{2^9}$$

and that the range of the counter is from $T \times fmin$ to $T \times fmax$, or a range of $T \times (fmax - fmin)$ from zero to full scale. The resolution of each counter bit is then

$$LSB = \frac{T \times (fmax - fmin)}{2^n}$$

where n is the desired number of bits to be resolved. For our example, $T = 512/5000 = .1024$ seconds; .1 second yields a slightly better accuracy.

From earlier tries at generating decimal time delays in Chapter 7, it has been amply demonstrated that these cannot be done perfectly using a 16 megahertz crystal (.75 microsecond count interval). We will be close enough to meet our requirements.

T1 is used in the auto-reload mode 2 to generate overflow interrupts every 192 microseconds ($256 \times .75$ microseconds). These overflows are counted using R4 and R5 until .100032 seconds have elapsed (521d overflows). For this example, T0 is used as a counter to count the external frequency that is fed to the port 3.4 (T0) pin during the T1 interval. Using the interval chosen, the range of counts in T0 becomes

$$0V = 1000 \text{ Hz} \times .100032 \text{ s} = 100d \text{ counts}$$
$$5V = 6000 \text{ Hz} \times .100032 \text{ s} = 600d \text{ counts}$$
$$.01V = 10Hz \times .100032 \text{ s} = 1 \text{ count}$$

which meets the desired accuracy specification.

Freq

The program "freq" uses T0 to count an external pulse train that is known to vary in frequency from 1000 to 6000 hertz. T1 generates an exact time delay of 192 microseconds that is counted using registers R4 and R5 of bank 1 until T1 has overflowed 521d times, or a total delay of .100032 seconds.

ADDRESS	MNEMONIC	COMMENT

```
              .equ frqflg,0fh      ;use addressable bit for a flag
              .org 0000h
freq:         mov sp,#0fh          ;set stack above register bank one
              sjmp over            ;jump over the T1 interrupt location
;
;T1 will overflow and vector here; R4 and R5 will be used as a
;combined 16-bit counter to count the 521d overflows; the extra
;microseconds needed to detect end of count and stop T0 will
;introduce a slight error
;
              .org 001bh           ;place program at T1 interrupt vector
              setb psw.3           ;switch to register bank 1
              djnz r4,timup        ;count R4 down and test for 521d
                                   ;counts
              dec r5
timup:        cjne r5,#0fdh,go     ;count down from 0000 to FDF7h (209h)
              cjne r4,#0f7h,go     ;209h = 521d
              clr tcon.4           ;stop T0 and set frqflg
              setb frqflg          ;main program can now process
                                   ;frequency
              clr tcon.6           ;stop T1
go:           clr psw.3            ;return to register bank zero
              reti                 ;total extra time to stop T0 =
                                   ;8.25 µs
;
;the main program sets up T0 to be a counter and starts T1; the
;flag frqflg is then watched until it is set by the interrupt
;program; the main program must do this every time a frequency read
;is desired; if continuous frequency determinations are desired by
;the main program, then the interrupt program could call a subroutine
;frequency handling program inserted before ''go'' in place of the
;instruction that stops T1.
;
over:         setb psw.3           ;select register bank one
              mov r4,#00h          ;reset R4 and R5
              mov r5,#00h
              clr psw.3            ;restore to register bank zero
              mov tmod,#25h        ;T1 mode 2 timer, T0 mode 1 counter
              mov tl1,#00h         ;count up from 00 and reset
              mov th1,#00h         ;reload with 00
              mov tcon,#50h        ;start T0 and T1
              mov ie,#88h          ;enable T1 to interrupt
simulate:     jbc frqflg,getfrq    ;simulate main program getting data
              sjmp simulate
getfrq:       nop                  ;place frequency subroutine here
              sjmp simulate
              .end
```

◁ **COMMENT** ───

The longer the time taken to count, the more accurate the frequency will be (but remember, it makes little sense to make the readout more accurate than the basic sensor). T0 will overflow at 65,535 or at the end of an interval of 10.92 s at fmax, which can be generated in T1 and R4, R5. In this case, the accuracy would be to the nearest .09 hertz (.0001 volt).

If you wish to generate a delay closer to .1 s than used in the example, make T1 cycle in a shorter period of time and count these shorter periods in R4, R5. Compensate for the 8.5 micro-seconds it takes for the interrupt routine to determine that time is up.

Preloading T0 with a number that causes T0 to overflow to 0000 when fmin is present during T will enable T0 to read the voltage directly. For our example, presetting T0 to FF9Ch will have T0 = 01F4h (500d) at fmax = 60,000 hertz for T = .1 s.

Pulse Width Measurement

Theoretically, if the input pulse is known to be a perfect square wave, the pulse frequency can be measured by finding the time the wave is high (Th). The frequency is then

$$UF = \frac{1}{Th \times 2}$$

If Th is 200 microseconds, for example, then UF is 2500 hertz. The accuracy of the measurement will fall as the input wave departs from a 50 percent duty cycle.

Timer X may be configured so that the internal clock is counted only when the corresponding $\overline{INTX}$ pin is high by setting the GATE X bit in TMOD. The accuracy of the measurement is within approximately one-half of the timer clock period, or .375 microsecond for a 16 megahertz crystal. This accuracy can only be attained if the measurement is started when the input wave is low and stopped when the input next goes low. Pulse widths greater than the capacity of the counter, which is 49.152 milliseconds for a 16 megahertz crystal, can be measured by counting the overflows of the timer flag and adding the final contents in the counter.

For the example in this section, the sensor used to measure the 0 volt to 5 volts dc voltage has a fixed frequency of 1000 hertz or a period of 1 ms. For a 0 volt input, the sensor is high for 400 microseconds and low for 600 microseconds; when the sensor input is 5 volts, the output is high for 900 microseconds and low for 100 microseconds. Each volt represents 100 microseconds of time; the accuracy of the measurement is ±.00325 volts, which is within the specification of .01 volt.

To make the measurement, T0 will be configured to count the internal clock when $\overline{INT0}$ is high. The measurement is not started until $\overline{INT0}$ goes from high to low, leaving a minimum of 100 microseconds to start T0. The measurement is made while $\overline{INT0}$ is high and stopped when $\overline{INT0}$ goes low again. The whole process can be interrupt driven by using the interrupt flag associated with $\overline{INT0}$. The IE0 flag can be set whenever $\overline{INT0}$ goes from high to low to notify the program to start the pulse width timing and then to stop. A variation of this program is currently in use to measure fabric width by measuring the reflection time of a scanning laser.

Width

The program "Width" measures the width of pulses that are fed to the $\overline{INT0}$ pin, port 3.2 and that are known to vary from 400 to 900 microseconds. The program starts when the interrupt flag IE0 is set and stops the next time the flag is set, indicating one complete cycle of the input wave.

ADDRESS	MNEMONIC	COMMENT

```
                .equ wflg,00h      ;flag set to notify main program
                .org 0000h
width:          sjump over         ;jump over INTO flag vector location
;
;the INTO edge triggered flag will vector here
;
                .org 0003h
                jbc tcon.4,stop    ;if T0 is running, stop T0
                setb tcon.4        ;if T0 is not running, enable T0
                clr wflg           ;reset wflg until next measurement
                reti               ;return with T0 enabled
stop:           setb wflg          ;set flag for main program
                reti               ;return with T0 stopped
;
;the main program resumes here; the program monitors the flag that
;indicates that a width measurement has just been made
;
over:           ;mov tmod,#09h     ;set T0 to count when INTO high
                mov tcon,#01h      ;enable edge trigger for INTO
                mov tl0,#00h       ;reset T0
                mov th0,#00h
                mov ie,#81h        ;enable external interrupt
simulate:       jbc wflg,getw      ;look for wflg and get width
                sjmp simulate
getw:           nop                ;real program would read T0 for width
                mov tl0,#00h       ;reset T0
                mov th0,#00h
                sjmp simulate      ;simulate main program
                .end
```

 COMMENT

If there is a considerable amount of electrical noise present on the INTO pin, an average value of the pulse width could be found by measuring the widths of a number of consecutive pulses. A counter could be incremented at the end of each cycle and the sum of the widths divided by the counter contents. The noise should average to zero.

Frequency can be measured by timing the interval of a number (M) of high-to-low INTX interrupts. Synchronize the timing by starting the timer at the first transition, and stop the timer at the Mth + 1 transition. The frequency is then

$$UF = \frac{M}{T}$$

where T is the count in the timer.

D/A and A/D Conversions

Conversion between the analog and digital worlds requires the use of integrated circuits that have been designed to interface with computers. Highly intelligent converters are commercially available that all have the following essential characteristics:

Parallel data bus: tri-state, 8-bit

Control bus: enable (chip select), read/write, ready/busy

The choice the designer must make is whether to use the converter as a RAM memory location connected to the memory busses or as an I/O device connected to the ports. Once that choice is made, the set of instructions available to the programmer becomes limited. The memory location assignment is the most restrictive, having only MOVX available. The design could use the additional 32K RAM address space with the addition of circuitry for A15. By enabling the RAM when A15 is low, and the converter when A15 is high, the designer could use the upper 32K RAM address space for the converter, as was done to expand port capacity by memory mapping in Chapter 7. All of the examples examined here are connected to the ports.

D/A Conversions

A generic R-2R type D/A converter, based on several commercial models, is connected to ports 1 and 3 as shown in Figure 8.10. Port 1 furnishes the digital byte to be converted to an analog voltage; port 3 controls the conversion process. The converter has these features:

$Vout = -Vref \times (byte\ in/100II)$, $Vref = \pm 10\ V$

Conversion time: 5 μs

Control sequence: $\overline{CS}$ then $\overline{WR}$

For this example, a 1000 hertz sine wave that will be generated can have a programmable frequency. Vref is chosen to be -10 volts, and the wave will swing from $+9.96$ volts to 0 volt around a midpoint of 4.48 volts. The program uses a lookup table to generate the amplitude of each point of the sine wave; the time interval at which the converter is fed bytes from the table determines the wave frequency.

The conversion time limits the highest frequency that can be generated using S sample point. In this example, the shortest period that can be used is

$$Tmin = S \times 5\ \mu s = 5S\ \mu s, \qquad fmax = \frac{200,000}{S}$$

FIGURE 8.10 D/A Converter Circuit for "Davcon" Program

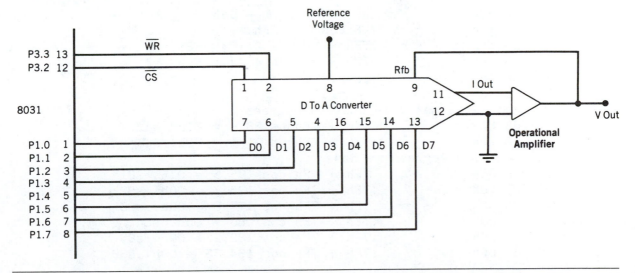

The design tension is high frequency versus high resolution. For a 1000 hertz wave, S could be 200d samples. In reality, we cannot use this many samples; the program cannot fetch the data, latch it to port 1, and strobe port 3.3 in 5 microseconds. An inspection of the program will show that the time needed for a single wave point is 6 microseconds, and setting up for the next wave takes another 2.25 microseconds. S becomes 166d samples using the 6 microseconds interval, and the addition of 2.25 microseconds at the end of every wave yields a true frequency of 1001.75 hertz.

Davcon

The D/A converter program "Davcon" generates a 1000 hertz sine wave using an 8-bit converter. 166d samples are stored in a lookup table and fed to the converter at a rate of one sample every 6 microseconds. The lookup table is pointed to in external ROM by the DPTR, and R1 is used to count the samples. Numbers in parentheses indicate the number of cycles.

ADDRESS	MNEMONIC	COMMENT
	.org 0000h	
davcon:	clr p3.2	;enable chip select to converter
	mov dptr,#table	;get base address to DPTR
repeat:	mov r1,#0a6h	;initialize R1 to 166d (1)
next:	mov a,r1	;offset into table (1)
	movc a,@a+dptr	;get sample (2)
	mov p1,a	;sample to port 1 (1)
	clr p3.3	;write strobe low (1)
	setb p3.3	;write strobe high (1)
	djnz r1,next	;loop for 166D samples (2)
	sjmp repeat	;reload R1 and generate next wave (2)

```
;
;the lookup table begins here; a cosine wave is chosen to make the
;table readable; the first 83 samples cover the wave from maximum to
;1 less than 0; the next 83 cover the wave from 0 to maximum. 83
;samples per half-cycle means a sample every 2.17 degrees
;
table:     .db 00h              ;no entry at A = 00h
           .db ffh              ;FFhcos 0 = FFh. s1
           .db feh              ;7Fh + 7Fhcos 2.17 = FEh. s2
           .db feh              ;7Fh + 7Fhcos 4.34 = FEh. s3
           .db fdh              ;sample 4
           .db fdh              ;sample 5
           ;and so on until we near 90 degrees:
           .db 81h              ;7Fh + 7Fhcos 88.9 = 81h. s42
           .db 7ch              ;7Fh + 7Fhcos 91.1 = 7Ch. s43
           ;near 180 degrees we have:
           .db 01h              ;7Fh + 7Fh cos 173.5 = 01h. s81
           .db 00h              ;7Fh + 7Fh cos 175.7 = 00h. s82
           .db 00h              ;7Fh + 7Fh cos 177.8 = 00h. s83
           .db 00h              ;7Fh + 7Fh cos 180 = 00h. s84.
           .db 00h              ;7Fh + 7Fh cos 182.2 = 00h. s85
           .db 00h              ;7Fh + 7Fh cos 184.33 = 00h. s86
```

Continued

ADDRESS	MNEMONIC	COMMENT
	.db 01h	;7Fh + 7Fh cos 186.5 = 01h. s87

;finally, close to 360 degrees the table contains:

	MNEMONIC	COMMENT
	.db fbh	;s 161
	.db fch	;s 162
	.db fdh	;s 163
	.db fdh	;s 164
	.db feh	;s 165
	.db feh	;s 166
	.end	

 COMMENT ─────────────────────────────────────

The program retrieves the data from the highest to the lowest address.

A/D Conversion

The easiest A/D converters to use are the "flash" types, which make conversions based upon an array of internal comparators. The conversion is very fast, typically in less than 1 microsecond. Thus, the converter can be told to start, and the digital equivalent of the input analog value will be read one or two instructions later. Modern successive approximation register (SAR) converters do not lag far behind, however, with conversion times in the 2–4 microsecond range for eight bits.

At this writing, flash converters are more expensive (by a factor of two) than the traditional SAR types, but this cost differential should disappear within four years. Typical features of an eight-bit flash converter are

Data: Vin = Vref(−), data = 00h; Vin = Vref(+), data = FFh

Conversion time: 1 μs

Control sequence: $\overline{CS}$ then $\overline{WR}$ then $\overline{RD}$

An example circuit, using a generic flash converter, is shown in Figure 8.11. Port 1 is used to read the byte value of the input analog voltage, and port 3 controls the conversion.

FIGURE 8.11   A/D Converter Circuit for "Adconv" Program

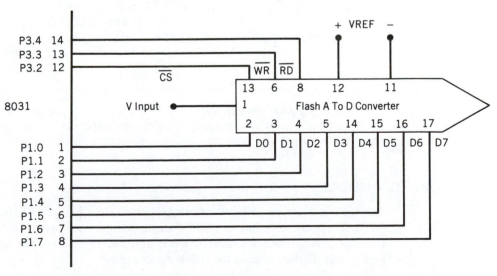

A conversion is started by pulsing the write line low, and the data is read by bringing the read line low.

Our example involves the digitizing of an input waveform every 100d microseconds until 1000d samples have been stored in external RAM.

Adconv

The program "Adconv" will digitize an input voltage by sampling the input every 100 μs and storing the digitized values in external RAM locations 4000h to 43E7h (1000d samples). Numbers in parentheses are cycles. The actual delay between samples is 99.75 microseconds.

ADDRESS	MNEMONIC	COMMENT
	.equ begin,4000h	;start storage at 4000h
	.equ delay,74h	;delay in DJNZ loop for 87 usec
	.equ end1,43h	;high byte of ending address
	.equ end2,e8h	;low byte of ending address
	.org 0000h	
adconv:	mov dptr,#begin	;point to starting address in RAM
	clr p3.2	;generate $\overline{CS}$ to ADC
next:	clr p3.3	;generate $\overline{WR}$ pulse (1)
	setb p3.3	;(1)
	clr p3.4	;generate $\overline{RD}$ pulse (1)
	mov a,pl	;get data (1)
	setb p3.4	;end of $\overline{RD}$ pulse (1)
	movx @dptr,a	;store in external RAM (2)
	inc dptr	;point to next and see if done (2)
	mov a,dph	;(1)
	cjne a,#end1,wait	;(2)
	mov a,dpl	;(1)
	cjne a,#end2,wait	;(2)
	sjmp done	;finished if both tests pass
wait:	mov r1,#delay	;delay for 87d μs
here:	djnz r1,here	;(2) × .75 μs × 116d = 87 μs
	sjmp next	;(2) 17d cycles (12.75 μs)
done:	sjmp done	;simulate rest of program
	.end	

COMMENT

Using this program, we could fill up the RAM in 3.2 s, which illustrates the volumes of data that can be gathered quickly by such a circuit. Realistic applications would feature some data reduction at the microcontroller before the reduced (massaged) data were relayed to a host computer.

Multiple Interrupts

The 8051 is equipped with two external interrupt input pins: $\overline{INT0}$ and $\overline{INT1}$ (P3.2 and P3.3). These are sufficient for small systems, but the need may arise for more than two interrupt points. There are many schemes available to multiply the number of interrupt points; they all depend upon the following strategies:

Connect the interrupt sources to a common line

Identify the interrupting source using software

Because the external interrupts are active low, the connections from the interrupt source to the $\overline{\text{INTX}}$ pin must use open-collector or tri-state devices.

An example of increasing the $\overline{\text{INT0}}$ from one to eight points is shown in Figure 8.12. Each source goes to active low when an interrupt is desired. A corresponding pin on port 1 receives the identity of the interrupter. Once the interrupt program has handled the interrupt situation, the interrupter must receive an acknowledgment so that the interrupt line for that source can be brought back to a high state. Port 3 pins 3.3, 3.4, and 3.5 supply, via a 3-to-8 decoder, the acknowledgment feedback signal to the proper interrupt source. The decoder is enabled by port pin 3.0.

Multiple and simultaneous interrupts can be handled by the program in as complex a manner as is desired. If there is no particular urgency attached to any of the interrupts then they can be handled as the port 1 pins are scanned sequentially for a low.

A simple priority system can be established whereby the most important interrupt sources are examined in the priority order, and the associated interrupt program is run until finished. An elaborate priority system involves ordering the priority of each source. The elaborate system acknowledges an interrupt immediately, thus resetting that source's interrupt line, and begins executing the particular interrupt program for that source. A new interrupt from a higher priority source forces the current interrupt program to be suspended and the new interrupter to be serviced.

To acknowledge the current interrupt in anticipation of another, it is necessary to also re-arm the $\overline{\text{INTX}}$ interrupt by issuing a "dummy" RETI instruction. The mechanism for accomplishing this task is illustrated in the program named "hipri." First, a low priority scheme is considered.

FIGURE 8.12 Multiple-Source Interrupt Circuit Used in "Lopri" and "Hipri" Programs

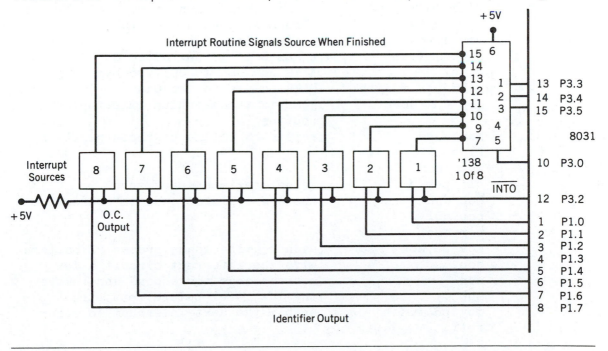

Lopri

The program "Lopri" scans port P1 for the source of an interrupt that has triggered $\overline{INT0}$. The pins are scanned for a low and the scan resumed after any interrupt is found and serviced. The interrupt source is acknowledged prior to a RETI instruction. R5 of bank 1 is used to store the next pin to be scanned, and R6 is used to scan the pins for a low. A jump table is used to select the interrupt routine that matches the particular interrupt. Each interrupt routine supplies the 3-to-8 decoder a unique acknowledge pattern before a RETI.

ADDRESS	MNEMONIC	COMMENT
	.equ ack,70h	;each interrupt routine loads its
		;unique acknowledge byte in ack
	.org 0000h	
lopri:	sjmp over	;jump over the $\overline{INT0}$ interrupt address

;
;The $\overline{INT0}$ interrupt will vector the program here
;

	.org 0003h	;$\overline{INT0}$ vector address
	mov ack,#0ffh	;place enable pattern in ack
	push acc	;save A
	push dpl	;save DPTR
	push dph	
	setb psw.3	;select register bank one
	mov a,r5	;get pattern in R5 to A
	orl a,p1	;OR the single zero in A with P1
	mov r6,#08h	;rotate A through C eight times
which:	rrc a	;find the zero starting at P1.0
	jnc low	;keep rotating until low found
	djnz r6,which	;if not found then it was not this pin
	sjmp goback	;return with no action taken
low:	mov a,r6	;convert from 1-of-8 low to number
	subb a,#01h	;A was 8 to 1, now 07 to 00
	rl a	;A is now 0Eh to 00 (two bytes/sjmp)
	mov dptr,#jmptbl	;DPTR points to the base of jump table
	jmp @a+dptr	;jump to the matching interrupt
		;routine
goback:	mov a,r5	;rotate r5 to the next pin position
	rl a	
	mov r5,a	
	clr psw.3	;select register bank zero
	pop dph	;restore register used in subroutine
	pop dpl	
	pop acc	
	mov p3,ack	;each routine loads proper P3 pattern
	nop	;give the interrupt circuit a few
	nop	;microseconds to respond and remove
	nop	;the low level before returning
	mov p3,#0ffh	;enable the next interrupt to occur
	reti	

Continued

ADDRESS	MNEMONIC	COMMENT

```
;
;the main program starts here followed by the interrupt routine jump
;table (simulated)
;
over:       mov sp,#0fh          ;move stack above register bank one
            mov p3,#0ffh         ;set port 3 to disable 3/8 all high
            setb psw.3           ;select bank one and set R5 to one low
            mov r5,#0feh         ;port pin 1.0 selected
            clr psw.3            ;return to bank zero
            mov ie,#81h          ;enable INTO interrupt
            mov tcon,#00h        ;enable level trigger for INTO
simulate:   sjmp simulate        ;simulate main program
jmptbl:     sjmp goback          ;simulate interrupt programs; pin 7
            sjmp goback          ;6
            sjmp goback          ;5
            sjmp goback          ;4
            sjmp goback          ;3
            sjmp goback          ;2
            sjmp goback          ;1
            sjmp goback          ;0
            .end
```

────▷── COMMENT ──

The instruction JMP @A+DPTR has been used to select one of a number of jump addresses, depending upon the number found in A. The simulated subroutines could be an SJMP to the actual interrupt handling subroutine. Because each SJMP takes two bytes to execute, A has to be doubled to point to every other byte in the jump table. When this action is not convenient, A can use a lookup table to get a new A, which then accesses a jump address.

R5 has one bit low, and that bit acts as a mask when ORed with P1 to find the low bit in P1. When the low pin does not match the R5 pattern, the RETI will immediately cause INTO to interrupt again, and R5 will be set to the next pin position. The worst-case response time, if eight pins must be searched before the low pin is found, will be in the order of 600 microseconds.

If INTO is triggered by noise, the routine returns after the first fruitless search with no action taken and re-arms the interrupt structure.

The external interrupt flags are cleared when the program vectors to the interrupt address *only* when the external interrupt is *edge* triggered. *Level* triggered interrupts must have the low level removed *before* the RETI, or an immediate interrupt is regenerated. Each interrupt routine loads the internal RAM location "ack" with the proper bit pattern to the decoder to enable and decode the proper line to reset the interrupting source.

Hipri

Suppose that we wish to have a priority system by which the priority of each input pin is assigned at a different level—that is, there are eight priority levels, and each higher level can interrupt one at a lower level. Theoretically, this leads to at least nine return addresses being pushed on the stack (plus any other registers saved), so the stack should be expected to grow more than 18d bytes; it is set above the addressable bits at location 2Fh.

In order to enable the interrupt structure in anticipation of a higher level interrupt, it is necessary to issue a RETI instruction without actually leaving the interrupt routine that currently has the highest priority. One way to accomplish this task is to push on the stack the address of the current interrupt routine to be done. Then, use a RETI that will return to the address on the stack, the desired current interrupt subroutine, and also re-arm the interrupt structure should another interrupt occur. The addresses of each subroutine can be known before assembly by originating each at a known address, or the program can find each address in a lookup table and push it on the stack, as illustrated in the example program.

For this example, the priority of each interrupt source is equivalent to the port 1 pin to which its identity line is connected. P1.0 has the highest priority, and P1.7, the lowest. A lookup table is used to find the address of the subroutine to be pushed on the stack.

External interrupt $\overline{INT0}$ is connected to the common interrupt line from all sources. It is enabled edge triggered whenever an interrupt routine is running so that any higher priority interrupt will be immediately acknowledged. If a lower priority interrupt occurs, it will interrupt the program in progress long enough to determine the priority. The interrupted subroutine will resume, and the lower level interrupt source priority will be saved until the subroutine in progress is finished. All interrupting sources maintain their identity lines low until they are acknowledged. The common interrupt line is reset immediately to enable any other source to interrupt the 8051.

If a higher level source interrupts a lower priority interrupt, then the high priority routine will interrupt the lower priority routine. The priority of the lower level interrupt will be saved.

The program "Hipri" assigns eight levels of priority to the interrupt sources connected to port 1. A lookup table is used to find the address of the interrupt handling subroutine that is pushed on the stack. A RETI instruction is then used to "return" to the desired subroutine and re-arm the interrupt hardware on the 8051.

ADDRESS	MNEMONIC	COMMENT
	.org 0000h	
hipri:	ljmp over	;jump over the $\overline{INT0}$ routine
;the $\overline{INT0}$ interrupt will vector here to find the identity and		
;priority of the interrupt source		
;		
	.org 0003h	;$\overline{INT0}$ interrupt vectors here
int:	push dph	;save registers used
	push dpl	
	push acc	
	setb psw.3	;use register bank one
	clr p3.0	;reset common $\overline{INT}$ line by strobing
	setb p3.0	;pin 3.0
	mov dptr,#base	;get base address of address table
	mov a,R5	;get priority of current interrupt
	orl a,P1	;determine if new interrupt is ;higher
	cjne a,#0ffh,higher	;A will be FFh if new < old
	pop acc	;not higher priority; return to ;current
	pop dpl	
	pop dph	
	reti	

Continued

ADDRESS	MNEMONIC	COMMENT

```
higher:     push 0dh              ;higher priority; save old (R5)
            jnb acc.0,first       ;find higher priority interrupt
            jnb acc.1,second
            jnb acc.2,third
            jnb acc.3,fourth
            jnb acc.4,fifth
            jnb acc.5,sixth
            jnb acc.6,seventh
            jnb acc.7,eighth
            sjmp goback           ;noise; return with no new interrupt
first:      mov r5,#0ffh          ;highest priority
            mov a,#00h            ;load A with offset into lookup
                                  ;table
            sjmp pushadd          ;"pushadd" will push the address
second:     mov r5,#0feh          ;may only be interrupted by P1.0
            mov a,#02h            ;load A with offset for next program
            sjmp pushadd
third:      mov r5,#0fch          ;interrupt by 0-1
            mov a,#04h
            sjmp pushadd
fourth:     mov r5,#0f8h          ;interrupt by 0-2
            mov a,#06h
            sjmp pushadd
fifth:      mov r5,#0f0h          ;interrupt by 0-3
            mov a,#08h
            sjmp pushadd
sixth:      mov r5,#0e0h          ;interrupt by 0-4
            mov a,#0ah
            sjmp pushadd
seventh:    mov r5,#0c0h          ;interrupt by 0-5
            mov a,#0ch
            sjmp pushadd
eighth:     mov r5,#80h           ;interrupt by 0-6
            mov a,#0eh
pushadd:    mov r6,a              ;save A for second byte fetch
            inc a                 ;point to the low byte of the
                                  ;address
            movc a,@a+dptr        ;get first program address low byte
            push acc              ;push the low byte
            mov a,r6              ;get A back
            movc a,@a+dptr        ;get the high byte of the address
            push acc              ;push the high byte
            reti                  ;execute subroutine; enable
                                  ;interrupt
goback:     pop 0dh               ;restore old priority mask
            mov a,pl              ;look at P1 for more interrupts
            cjne a,#0ffh,old      ;see if any are waiting, or in
                                  ;progress
```

Continued

ADDRESS	MNEMONIC	COMMENT

Continued

```
                pop acc              ;if none waiting then return to main
                pop dpl              ;program
                pop dph
                clr psw.3            ;return to register bank 0
                reti                 ;return to main program
old:            orl a,r5             ;A = FFh if next interrupt
                                     ;waiting was
                cjne a,#0ffh,next    ;itself interrupted
                pop acc              ;get old interrupt values
                pop dpl
                pop dph
                reti                 ;return to old interrupt in progress
next:           pop acc              ;the waiting interrupt is a new one
                pop dpl              ;that has never begun to execute
                pop dph              ;jump to ''int'' as if an INTO has
                ljmp int             ;occured
;
;the lookup table that contains the addresses of the eight interrupt
;programs is assembled here; the assembler knows all the actual
;numbers at assembly time
;
base:           .dw prog1            ;progx is the actual interrupt
                                     ;routine
                .dw prog2
                .dw prog3
                .dw prog4
                .dw prog5
                .dw prog6
                .dw prog7
                .dw prog8
prog1:          nop                  ;simulate interrupt program.
                ljmp goback          ;be sure to acknowledge before ljmp
prog2:          nop                  ;after subroutine has finished
                ljmp goback
prog3:          nop
                ljmp goback
prog4:          nop
                ljmp goback
prog5:          nop
                ljmp goback
prog6:          nop
                ljmp goback
prog7:          nop
                ljmp goback
prog8:          nop
                ljmp goback
```

Continued

ADDRESS	MNEMONIC	COMMENT

```
;
;the main program starts here; "progx" could have been assembled
;after the main program if desired
;
over:      mov sp,#2fh          ;set stack above addressable bits
           setb tcon.0          ;enable INT0 edge-triggered
           setb psw.3           ;choose register bank one
           mov r5,#00h          ;set for interrupt at all levels
           mov ie,#81h          ;enable INT0
here:      clr psw.3            ;return to bank zero
           sjmp here            ;simulate main program
           .end
```

▷ COMMENT ───

The .dw assembler directive will store the *high* byte of the two-byte word at the *lower* address in memory. For the RETI in "pushadd" to work properly, the *low* address byte must be placed on the stack first.

If interrupt A has just gone low, and interrupt B, which is of a higher priority, occurs after the system has vectored to the INT0 address, interrupt B will be accessed if the B line goes low before the polling software starts (JNB ACC.x). If the polling has caused A to be chosen, then B will be recognized after the RETI in "pushadd" causes the A address to be POPed from the stack. One instruction of A will be executed, then the IE0 flag in TCON will cause an interrupt.

The 8051 interrupt system will generate an interrupt unless any of the following conditions are true:

Another routine of equal or greater priority is running.

The current instruction is not finished.

The instruction is a RETI or any IE/IP access.

The edge-triggered interrupt sets the IE0 flag, and the interrupt that generated the edge serviced after any of the listed conditions are cleared.

Hardware Circuits for Multiple Interrupts

Solutions to the expanded interrupt problem proposed to this point have emphasized using a minimal amount of external circuitry to handle multiple, overlapping interrupts. A hardware strategy, which can be expanded to cover up to 256 interrupt sources, is shown in Figure 8.13. This circuit is a version of the "daisy chain" approach, which has long been popular.

The overall philosophy of the design is as follows:

1. The most important interrupt source is physically connected first in the chain, with those of lesser importance next in line. Lower priority interrupt sources are "behind" (connected further from INT0) those of a higher priority.

2. Each interrupting source can disable all signals from sources that are wired behind it. All sources that lose the INACTOUT signal (a low level) from the source(s) ahead of it will place their source address buffer in a tri-state mode until INACTOUT is restored.

FIGURE 8.13 Daisy Chain Circuit Used for "Hardint"

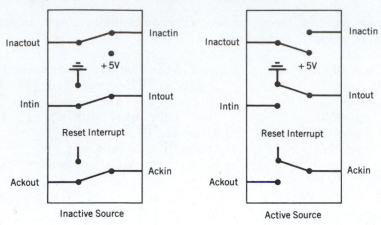

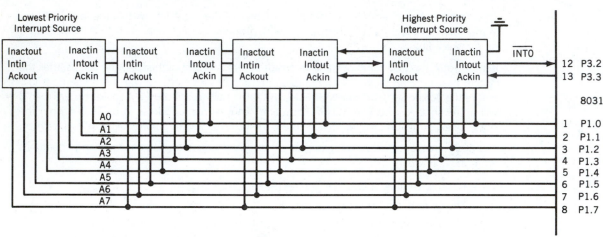

3. A requesting source pulls its $\overline{\text{INTOUT}}$ line low and places its 8-bit identifier on the tri-state bus connected to port 1. The interrupt routine at the $\overline{\text{INT0}}$ vector location reads P1 and, using a lookup table, finds the address of the subroutine that handles that interrupt. The address is placed on the stack and a RETI executed to go to that routine and re-arm the interrupt structure.

4. The interrupt subroutine generates an ACKIN signal (a low-level pulse) to the source from the 8051 at the end of the subroutine; the source then removes $\overline{\text{INTOUT}}$ and the 8-bit source address. When an interrupt is acknowledged, the interrupting source must bring the $\overline{\text{INTOUT}}$ line high for at least one machine cycle so that the 8051 interrupt structure can recognize the next high-to-low transition on $\overline{\text{INT0}}$.

The software is very simple for this scheme. Any interrupt received is always of higher priority than the one now running, and the source address on port 1 enables rapid access to the interrupt subroutine.

Accomplishing this interrupt sequence requires that the source circuitry be complex or that the source contain some intelligence such as might be provided by a microcontroller.

The additional source hardware will entail considerable relative expense for each source. As the number of interrupt sources increases, system costs rise rapidly. At some point the designer should consider another microcontroller that has extensive interrupt capability.

Hardint

The program "Hardint" is used with daisy-chained interrupt sources to service 16 interrupt sources. An interrupt is falling-edge triggered on $\overline{\text{INT0}}$ and the interrupt address read on P1. A lookup table then finds the address of the interrupt routine that is pushed on the stack and the RETI "returns" to the interrupt subroutine. The interrupt subroutine issues an acknowledgment on port 3.3, which resets the interrupting source.

ADDRESS	MNEMONIC	COMMENT

```
            .org 0000h
hardint:    ljmp over
;
;the interrupt program located at the INT0 vector address will read
;the source address on port 1, and push that address for a RETI to
;the interrupt subroutine for that address
;
            .org 0003h
            setb psw.3          ;choose register bank one
            push acc            ;save registers used
            push dpl
            push dph
            mov a,p1            ;read port 1 for source address
            cjne a,#10h,less    ;valid addresses are 00h to 0Fh
less:       jnc goback          ;invalid, return
            rl a                ;valid address, adjust A for addresses
            mov r6,a            ;save A for low byte fetch
            mov dptr,#base      ;point to program address lookup table
            inc a               ;point to low byte
            movc a,@a+dptr      ;get low byte
            push acc
            mov a,r6            ;get high byte offset
            movc a,@a+dptr
            push acc
            reti                ;execute subroutine; enable interrupt
goback:     pop dph             ;return to program in progress
            pop dpl
            pop acc
            clr psw.3           ;back to register bank zero
            reti
base:       .dw prog0           ;make lookup table for subroutines
            .dw prog1
            .dw prog2
            .dw prog3
            .dw prog4
            .dw prog5
```

Continued

ADDRESS	MNEMONIC	COMMENT

Continued

```
            .dw prog6
            .dw prog7
            .dw prog8
            .dw prog9
            .dw proga
            .dw progb
            .dw progc
            .dw progd
            .dw proge
            .dw progf
prog0:      nop                 ;simulate interrupt subroutine
            ljmp goback
prog1:      nop                 ;simulate interrupt subroutine
            ljmp goback
prog2:      nop                 ;simulate interrupt subroutine
            ljmp goback
prog3:      nop                 ;simulate interrupt subroutine
            ljmp goback
prog4:      nop                 ;simulate interrupt subroutine
            ljmp goback
prog5:      nop                 ;simulate interrupt subroutine
            ljmp goback
prog6:      nop                 ;simulate interrupt subroutine
            ljmp goback
prog7:      nop                 ;simulate interrupt subroutine
            ljmp goback
prog8:      nop                 ;simulate interrupt subroutine
            ljmp goback
prog9:      nop                 ;simulate interrupt subroutine
            ljmp goback
proga:      nop                 ;simulate interrupt subroutine
            ljmp goback
progb:      nop                 ;simulate interrupt subroutine
            ljmp goback
progc:      nop                 ;simulate interrupt subroutine
            ljmp goback
progd:      nop                 ;simulate interrupt subroutine
            ljmp goback
proge:      nop                 ;simulate interrupt subroutine
            ljmp goback
progf:      nop                 ;simulate interrupt subroutine
            ljmp goback
;
;place the main routine here
;
```

Continued

ADDRESS	MNEMONIC	COMMENT
over:	mov sp,#2fh	;set <u>stack</u> above addressable bits
	setb tcon.0	;set $\overline{INT0}$ for edge triggered
	mov ie,#81h	;enable $\overline{INT0}$
here:	sjmp here	;simulate main program
	.end	

COMMENT

If the lookup table goes beyond 128 addresses, or 256 bytes, then DPH is incremented by one to point to a second complete table.

Each interrupt subroutine must contain an acknowledge byte that is placed on P3 to reset each source.

Note the use of CJNE and the carry flag to determine relative sizes of two bytes at label "less."

Putting it all Together

All of the examples presented to this point have used the free ports (P1 and parts of P3) that the "cheap" design affords. It is clear that to do a real-world design requires the use of additional port chips to enable several functions to be interfaced to the 8051 at one time. Such a design is illustrated in this section, using an 8255 programmable port chip memory mapped at external RAM location 8000h to 8003h. A review of memory mapping found in Chapter 7 shows that the required address decoding can be done using an inverter to enable external RAM whenever A15 is low, and the 8255 whenever A15 is high. Actually, *any* address that begins with A15 high can address the 8255; 8000h seems convenient.

Ant

The example program uses the intelligent LCD display, a coded 16-key keypad, and is capable of serial data communications. This type of design is suitable for many applications where a small, inexpensive, alphanumeric terminal (dubbed the ANT) is needed for the factory floor or the student lab.

The design is shown in Figure 8.14. Port A of the 8255 is connected to the keypad. Port B supplies data bytes to the LCD and the lower half of port C controls the display. The program is interrupt driven by the keypad and the serial port. $\overline{INT0}$ is used to detect a keypress via the AND gate array while the serial interrupt is internal to the 8051. The serial port has the highest priority. This type of program is often called "multi-tasking" because the routines are called by the interrupt structure, and the computer appears to be doing many things simultaneously.

A keypad program developed in this chapter combined with a serial communication program from Chapter 9 completes the design.

The program "Ant" controls the actions of an 8051 configured as a terminal with a LCD display and hexadecimal keypad. The serial port is enabled, and has the highest priority of any function. The coded keyboard is a two-of-eight type which can use a lookup table to detect valid key presses. A shift key capability is possible because unique patterns are possible if one key is held down while another is pressed.

FIGURE 8.14   A Multi-Tasking Circuit Using Memory-mapped I/O

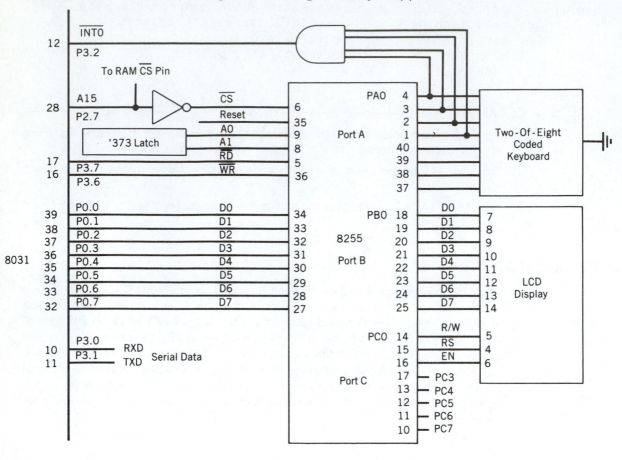

8255 I/O Ports Mapped At 8000, 8001, And 8002

ADDRESS	MNEMONIC	COMMENT
	.equ con,8003h	;address of 8255 mode control register
	.equ prta,8000h	;address of 8255 port A
	.equ prtb,8001h	;address of 8255 port B
	.equ prtc,8002h	;address of 8255 port C
	.equ conant,98h	;A = input, B and lower C = output
	.equ bf,0f3h	;C pattern to read LCD busy flag
	.equ wrd,0f2h	;C pattern to write data to LCD
	.equ wrc,0f0h	;C pattern to write control to LCD
	.equ setlcd,3fh	;initialize LCD to 2 lines, 5 × 10 dots
	.equ curs,06h	;LCD cursor blinks, moves left
	.equ lcdon,0eh	;LCD on

Continued

ADDRESS	MNEMONIC	COMMENT

```
                .equ klr,01h        ;clear LCD and home cursor
                .org 0000h
ant:            ljmp over           ;jump over the interrupt locations
;
;when a key is pressed, or a serial data character is sent or
;received, the program vectors to the interrupt address locations;
;dummy routines will be written here; refer to the key routines in
;this chapter and the serial data routines in Chapter 9 for examples
;of these programs
;
                .org 0003h          ;origin the keypad program here
                sjmp keypad         ;jump to keypad handling program
                .org 0023h          ;origin serial interrupt program here
                sjmp serial
keypad:         push dph            ;dummy keypad program, get the key
                push dpl
                mov dptr,#prta      ;read the key value
                movx a,@dptr        ;insert a key handling routine next
                pop dpl
                pop dph
                reti
serial:         nop                 ;dummy serial program
                reti
;
;the main program begins here; All the interrupts are initialized,
;the main program sends a ''hello'' to the display and waits for an
;interrupt
;
over:           setb tcon.0         ;set INT0 for edge triggered operation
                setb ip.4           ;set serial interrupt high priority
                acall serset        ;call the serial port setup routine
                mov dptr,#con       ;initialize 8255 mode to basic I/O
                mov a,#conant       ;set A = input, B and C = output
                movx @dptr,a        ;initialize 8255 mode register
                mov a,#setlcd       ;initialize the LCD and say ''hello''
                lcall lcdcon
                mov a,#curs
                lcall lcdcon
                mov a,#lcdon
                lcall lcdcon
                mov a,#klr
                lcall lcdcon        ;LCD is now initialzed and blank
                mov dptr,#msg       ;use DPTR to point to ''hello''
                lcall lcddta        ;send the message to the LCD
                mov ie,#91h         ;enable serial and INT0 interrupts
here:           sjmp here           ;simulate the rest of the program
serset:         ret                 ;dummy serial setup routine
```

Continued

ADDRESS   MNEMONIC   COMMENT
Continued

```
;
;The subroutine lcddta sends data characters to the LCD until the
;character ~ is found; the beginning of the message is passed to the
;subroutine in the DPTR by the calling program
;
lcddta:     mov a, #00h         ;get character of message
            movc a, @a+dptr
            cjne a, #7eh,mod    ;stop when ~ (7Eh) is found
            ret                 ;message sent
mod:        acall data          ;send data character
            inc dptr
            sjmp lcddta         ;loop until done
;
;the subroutine that sends control characters passed in A to the LCD
;display, via the 8255
;
lcdcon:     push dph            ;save registers used
            push dpl
            mov dptr,#prtb      ;get control data in A to port B
            movx @dptr,a
            mov dptr,#prtc      ;point to port C for LCD control
            mov a,#wrc          ;strobe character to LCD using port C
            movx @dptr,a
            mov a,#0ffh         ;end strobe
            movx @dptr,a
            acall dun           ;wait for LCD to finish
            pop dpl             ;restore registers
            pop dph
            ret
;
;the subroutine "data" sends data characters passed in A to the
;LCD screen for display
;
data:       push dph            ;save registers used
            push dpl
            mov dptr,#prtb      ;get character data in A to port B
            movx @dptr,a
            mov dptr,#prtc
            mov a,#wrd          ;strobe character to LCD using port C
            movx @dptr,a
            mov a,#0ffh         ;end strobe
            movx @dptr,a
            acall dun           ;wait for LCD to finish
            pop dpl             ;restore registers
            pop dph
            ret
;
;"dun" reads the busy flag on the LCD and returns the flag is low
```

Continued

ADDRESS	MNEMONIC	COMMENT

```
;
dun:       mov dptr,#con      ;configure port B as an input
           mov a,#9ah
           movx @dptr,a
           mov dptr,#prtc     ;set port C for a read command
           mov a,#bf
           movx @dptr,a       ;send command to read flag
           mov dptr,#prtb     ;read port B
           movx a,@dptr       ;the busy flag is bit 7
           jnb acc.7,go       ;done when BF = 0
           mov dptr,#prtc     ;if still busy then read again
           mov a,#0ffh
           movx @dptr,a
           sjmp dun
go:        mov a,#0ffh        ;finished, remove strobe
           mov dptr,#prtc
           movx @dptr,a
           mov dptr,#con      ;reset port B as an output
           mov a,#98h
           movx @dptr,a
           ret
;the message ''hello'' is assembled here; a great number of
;messages, each with a unique label, can be sent in this way
;
msg: .db ''hello~''
```

 COMMENT

The LCD example shows the extensive use of the DPTR and MOVX command when dealing with a memory mapped external port.

Forgetting to terminate every message with a ~ results in a very confused LCD as the remainder of ROM is written to the LCD.

There will be no interference between any of these programs if the serial interrupts always have priority. Serial data is received as it occurs, and the keypad program and any messages to the LCD are suspended for the few microseconds it takes to read the serial port. The suspended programs can resume until the next serial character, which is normally an interval of one or more milliseconds.

Summary

Hardware designs and programs have been illustrated to solve several common application problems that are especially suitable for solution using a microcontroller. These hardware circuits are

Keyboards: Lead-per-key, X–Y matrix, coded

Displays: 7-segment LED, intelligent LCD

Pulse measurement: frequency, pulse width

Data converters: R/2R digital to analog, flash analog to digital

Interrupts: multi-source, daisy chain

Expanded 8051 system: memory-mapped I/O

The programs in this chapter interface the 8051 to these circuits. New programming concepts introduced are

Interrupt handling

Register bank switching in "Svnseg"

Jump tables in "Lopri"

Stack RETI in "Hipri"

Using CJNE for relative size in "Hardint"

Multitasking in "Ant"

These programs can be used as the kernels for more comprehensive applications.

Problems

1. List the most likely effects if a keyboard program does not accomplish the following:
   a. Debounce keys when pressed down
   b. Check for a valid key code
   c. Wait for all keys up before ending keyboard routine
   d. Debounce keys when released

2. A keyboard has two keys: run and stop. Write a program that is interrupt driven by these two keys using $\overline{\text{INT0}}$ for the run key, and $\overline{\text{INT1}}$ for the stop key. If run is selected, set pin P3.0 high; if stop is selected, set the pin low. Bounce time is 10 milliseconds for the keys.

3. Determine why it is important to employ some kind of debounce subroutine in a keyboard program, particularly for interrupt driven programs, even if keys with absolutely no bounce are used.

4. The lookup table used in the program "Codekey" is very inefficient, using 256 bytes to form a table for the valid keys and using an FFh in all other locations for invalid keys. Write a subroutine using a series of CJNE instructions that will obtain the same result.

5. Repeat problem 4 by converting the keycode number in A from the codes B7h–EEh to 00–09h. One way to do this is to convert the first and second nibbles to the following numbers and then adding the nibbles to get a unique number:

CHANGE

First Nibble	Second Nibble	Add Converted Nibbles	
E to 0	E to 0	EE to 00	DD to 05
D to 4	D to 1	ED to 01	DB to 06
B to 8	B to 2	EB to 02	D7 to 07
	7 to 3	E7 to 03	BE to 08
		DE to 04	BD to 09

Note: Lookup tables can be used for each conversion, with invalid codes in both nibble lookup tables set to return numbers that, when added, sum to greater than 09.

6. Write a lookup table subroutine for the program "Bigkey" that will convert the row and column bytes for each key to a single byte number.

7. Expand the lookup table "convert" in the program "Svnseg" to include these characters: G, H, I, J, L, O, P, S, T, and U.

8. Write a program that will display the following message on the intelligent display:

''Hello!
Please Enter Command.''

Center each line of the display.

9. Write a subroutine that is past the starting address of an ASCII string in ROM and then displays the string on the intelligent display. The string length is fixed.

10. Repeat Problem 9 for a string of any length.

11. Write a program for the LCD display that will display the contents of register R1 as follows:

R1 = XX

XX is the R1 contents in hex. Center the display. (Hint: Remember the contents are in hex, and the display speaks ASCII.)

12. Write a program using timer 0 that will delay exactly .100000 milliseconds ±1 microsecond from the time the timer starts until it is stopped. (The crystal frequency is 16 megahertz).

13. Make a table that shows the accuracy of pulse width measurements as a function of multiples of count periods (.75 microseconds). The table should be arranged as follows:

PULSE WIDTH (×.75 μs)	ACCURACY (%)
1	
2	
3	
4	
5	
6	
7	
8	
9	
10	
20	
50	
100	

14. Write a program that can use the stack to "return" to any of 256 subroutines pointed to by the number 00 to FFh in A.

15. Compose a 40-value lookup table that will generate a sawtooth wave using a D/A converter.

16. Repeat Problem 15 without using a lookup table of any kind.

17. Repeat Problem 15 for a rectified sine wave.

18. Outline a method of measuring the frequency of a sine wave using a flash A/D converter. Estimate the highest frequency that can be measured to an accuracy of 1 percent.

19. In the section on measuring frequency, an expression was found for n bit resolution of a frequency measured over time, T:

$$LSB = \frac{T \times (fmax - fmin)}{2^n}$$

Derive an equivalent expression for the resolution of a frequency to n bits by measuring the period of M of the cycles.

20. Write a program that finds frequency by measuring the time for M cycles of the unknown periodic wave. Estimate the highest frequency that can be measured to an accuracy of 1 percent if the crystal is 16 megahertz.

21. Write a program that performs all of the functions of an intelligent daisy chain interrupt source controller.

22. Write a lookup table program for the "Ant" program that will allow the F key of a two-of-eight coded keypad to be used as a shift key. A shift key makes possible 31 valid key combinations. The key codes are

KEY	1	2	3	4	5	6	7	8	9
0	x				x				x
1	x					x			x
2	x						x		x
3	x							x	x
4		x			x				x
5		x				x			x
6		x					x		x
7		x						x	x
8			x		x				x
9			x			x			x
A			x				x		x
B			x					x	x
C				x	x				x
D				x		x			x
E				x			x		x
F				x				x	x

(Header over columns 1–8: **OUTPUT PIN**)

x means a connection is made; pin 9 is the common pin for all codes.

CHAPTER

9

Serial Data Communication

Chapter Outline

Introduction

Network Configurations

8051 Data Communication Modes

Summary

Introduction

Chapter 2 contained an extensive review of serial data communication concepts and the hardware and software that is built into the 8051 for enabling serial data transfers. Chapter 7 contained some brief programming examples of how this capability may be used. Serial data transmission has become so important to the overall computing strategy of industrial and commercial applications that a separate chapter on this crucial subject is appropriate.

One hallmark of contemporary computer systems is interconnectivity: the joining of computers via data networks that link the computers to each other and to shared resources, such as disk drives, printers, and other I/O devices. The beginning of the "computer age" saw isolated CPUs connected to their peripherals using manufacturer-specific data transmission configurations. One of the peripherals, however, was the teletype that had been borrowed from the telephone industry for use as a human interface to the computer, using the built-in keyboard and printer.

The teletype was designed to communicate using standard voice grade telephone lines via a *modem* (Modulator demodulator) that converts digital signals to analog frequencies and analog frequencies to digital signals. The data, by the very nature of telephone voice transmission, is sent and received serially. Various computer manufacturers adapted their equipment to fit the teletype, and, perhaps, the first "standard" interface in the industry was born.

This standard was enhanced in the early 1960's with the establishment of an electrical/mechanical specification for serial data transmission that was assigned the number RS 232 by the Electronics Industry Association. A standard data code was also defined for all the characters in the alphabet, decimal numbers, punctuation marks, and control characters. Based on earlier telephonic codes, the standard became known as the American Standard Code for Information Interchange (ASCII).

The establishment of RS 232 and ASCII coincided with the development of multi-user computer organizations wherein a number of users were linked to a host mainframe via serial data links. By now, the CRT terminal had replaced the slower teletype, but the RS 232 serial plug remained, and serial data was encoded in ASCII. Peripheral devices, such as printers, adopted the same standards in order to access the growing market for serial devices.

Serial data transmission using ASCII became so universal that specialized integrated circuits, Universal Asynchronous Receiver Transmitters (UARTS) were developed to perform the tasks of converting an 8-bit parallel data byte to a 10-bit serial stream and converting 10-bit serial data to an 8-bit parallel byte. When the second-generation 8051 microcontroller was designed, the UART became part of the circuit.

Chapter 7 introduced the basic programming concepts concerning transmitting and receiving data using the serial port of the 8051. In this chapter, we study the serial data modes available to the programmer and develop programs that use these modes. The four modes are as follows:

Mode 0: Shift register mode

Mode 1: Standard UART mode

Mode 2: Multiprocessor fixed mode

Mode 3: Multiprocessor variable mode

In this chapter, we also identify multiprocessor configurations that are appropriate for each mode and write sample programs to enable data communication between 8051 microcontrollers.

Network Configurations

The first problem faced by the network system designer is how to physically hook the computers together. The two possible basic configurations are the star and the loop, which are shown in Figure 9.1.

The star features one line from a central computer to each remote computer, or from "host" to "node." This configuration is often used in time-sharing applications when a central mainframe computer is connected to remote terminals or personal computers using a dedicated line for each node. Each node sees only the data on its line; all communication is private from host to node.

The loop uses one communication line to connect all of the computers together. There may be a single host that controls all actions on the loop, or any computer may be enabled to be the host at any given time. The loop configuration is often used in data-gathering applications where the host periodically interrogates each node to collect the latest information about the monitored process. All nodes see all data; the communication is public between host and nodes.

Choosing the configuration to use depends upon many external factors that are often beyond the control of the system designer. Some general guidelines for selection are shown in the following table:

FIGURE 9.1 Communication Configurations

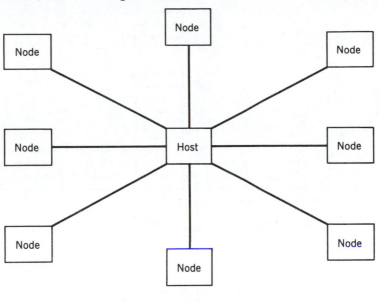

Star Configuration

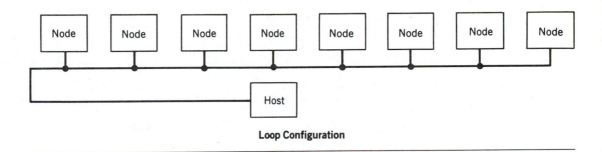

Loop Configuration

Objective	Network	Comments
Reliability	Star	Single node loss per line loss
Fault isolation	Star	Fault traceable to node and line
Speed	Star	Each node has complete line use
Cost	Loop	Single line for all nodes

The star is a good choice when the number of nodes is small, or the physical distance from host to node is short. But, as the number of nodes grows, the cost and physical space represented by the cables from host to nodes begins to represent the major cost item in the system budget. The loop configuration becomes attractive as cost constraints begin to out-weigh other considerations.

Microcontrollers are usually applied in industrial systems in large numbers distributed over long distances. Loop networks are advantageous in these situations, often with a host controlling data transmission on the loop. Host software is used to expedite fault isolation

FIGURE 9.2 Hybrid Communication Configurations

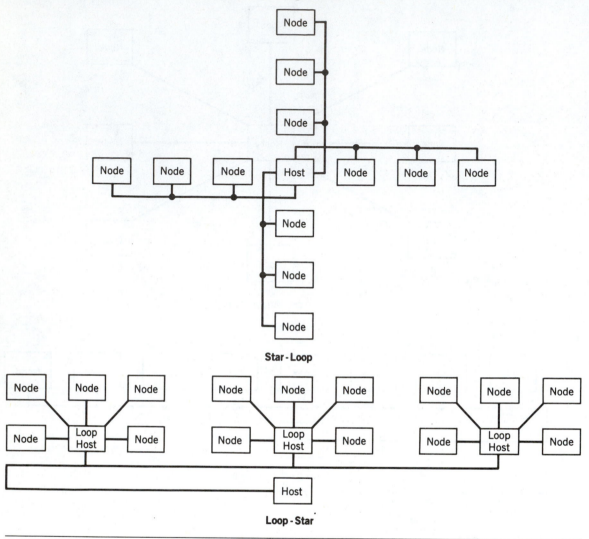

Star - Loop

Loop - Star

and, thus, improve system reliability. High speed data transmission schemes can be employed to enhance system response time where necessary.

The old racing adage "Speed costs money: How fast do you want to go?" should be kept in mind when designing a loop system. Successors to RS 232, most notably RS 485, have given the system designer 100 kilobaud rates over 4000-foot distances using inexpensive twisted-pair transmission lines. Faster data rates are possible at shorter distances, or more expensive transmission lines, such as coaxial cable, can be employed. Remember that wiring costs are often the major constraint in the design of large distributed systems.

Many hybrid network arrangements have evolved from the star and the loop. Figure 9.2 shows two of the more popular types that contain features found in both basic configurations.

8051 Data Communication Modes

The 8051 has one serial port—port pins 3.0 (RXD) and 3.1 (TXD)—that receives and transmits data. All data is transmitted or received in two registers with one name: SBUF. Writing to SBUF results in data transmission; reading SBUF accesses received data. Transmission and reception can take place simultaneously, and the receiver can be in the process of receiving a byte while a previous byte is still in SBUF. The first byte must be read before the reception is complete, or the second byte will be lost.

Physically the data is a series of voltage levels that are sampled, in the center of the bit period, at a frequency that is determined by the serial data mode and the program that controls that mode. All devices that wish to communicate must use the same voltage levels, mode, character code, and sampling frequency (baud rate). The wires that connect the ports must also have the same polarity so that the idle state, logic high, is seen by all ports.

The installation and checkout of a large distributed system are subject to violations of all of the "same" constraints listed previously. Careful planning is essential if cost and time overruns are to be avoided.

The four communication modes possible with the 8051 present the system designer and programmer with opportunities to conduct very sophisticated data communication networks.

Mode 0: Shift Register Mode

Mode 0 is not suitable for the interchange of data between 8051 microcontrollers. Mode 0 uses SBUF as an 8-bit shift register that transmits and receives data on port pin 3.0, while using pin 3.1 to output the shift clock. The data and the shift clock are synchronized using the six internal machine states, and even for microcontrollers using the same crystal frequency, they can be slightly out of phase due to differences in reset and start-up times.

Figure 9.3 shows the timing for the transmission and reception of a data character. Remember that the shift clock is generated internally and is always *from* the 8051 *to* the external shift register. The clock runs at the machine cycle frequency of f/12. Note that transmission is enabled any time SBUF is the destination of a write operation, regardless of the state of the transmitter empty flag, SCON bit 1 (TI).

FIGURE 9.3   Mode 0 Timing

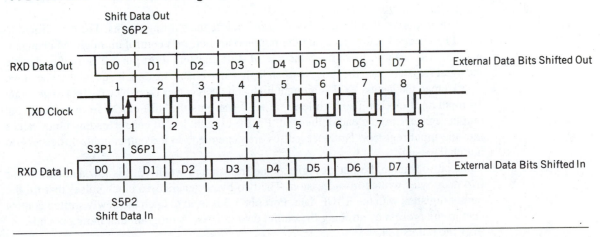

Data is transmitted, LSB first, when the program writes to SBUF. Data is shifted right during S6P2, or 24/f seconds after the rising edge of the shift clock at S6P1. Data is stable from just after S6P2 for one cycle. Good design practice dictates that the data be shifted into the external shift register during the high-to-low transition of the shift clock, at S3P1, to avoid problems with clock skew.

The receiver is enabled when SCON bit 5 (REN) is enabled by software and SCON bit 0 (RI) is set to 0. At the end of reception RI will set, inhibiting any form of character reception until reset by the software. The condition of RI cleared to 0 is unique for mode 0; all other modes are enabled to receive when REN is set without regard as to the state of RI. The reason is clear: Mode 0 is the only mode that controls when reception can take place. Enabling reception also enables the clock pulses that shift the received data into the receiver.

Reception begins, LSB first, with the data that is present during S5P2, or 24/f seconds before the rising edge of the shift clock at S6P1. The incoming data is shifted to the right. Incoming data should be stable during the low state of the shift clock, and good design practice indicates that the data be shifted from the external shift register during the low-to-high transition of the shift clock, at S6P1, so that the data is stable up to one clock period before it is sampled.

A serial data transmission interrupt is generated at the end of the transmission or reception of bit eight if enabled by the ES interrupt bit EI.4 of the enable interrupt register. Software must reset the interrupting bit RI or TI. As the same physical pin is used for transmission and reception, simultaneous interrupts are not possible.

Mode 0 is well suited for rapid data collection and control of multi-point systems that use a simple two-wire system for data interchange. Multiple external shift registers can expand the external points to an almost infinite number, limited only by the response time desired for the application. For instance, at f = 16 megahertz, each point of a 10,000 point system could be monitored every 60 milliseconds. Common industrial systems do not require rates this high, and a reasonable rate of one point per second would leave adequate time for processing by the program.

Modezero

A small system that features 16 points of monitored data and 16 points of control is shown in Figure 9.4. Data from the process is converted from parallel to serial in the '166 type registers. Data to the process is converted from serial to parallel in the type '164 registers and latched into the '373 latches.

It is important that the data be "frozen" before the shifting begins. The bits shifted in could be changed before reaching the microcontroller, or a control bit might be changed, momentarily, as it shifts through the output shift registers. Port pin 3.2 is used to disable the input registers from the process when high and to enable loading input values when low.

To read the inputs, P3.2 is brought high and the receiver is enabled (twice) to generate 16 input shift clocks. The high level on P3.2 prevents the shift clocks from reaching the output registers. At the end of the read, P3.2 is brought low to enable loading input values into the input registers. No clock pulses are generated, so the output control registers do not change state.

Control bits to the output registers are transmitted when P3.2 is low and SBUF has two data bytes written in succession. The two bytes generate 16 clock pulses that fill the output registers with the SBUF data. Port pin 3.3 is used to latch the newly shifted control data to the process by strobing the output data latches. A program that monitors and controls the points follows.

FIGURE 9.4   Shift Register Circuit Used with Modezero Program

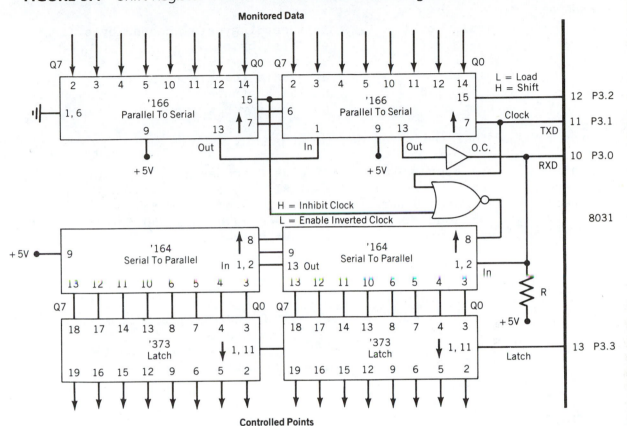

The program "modezero" monitors 16 bits and controls the state of 16 bits. The system can be expanded indefinitely by expanding the shift register configurations shown in Figure 9.4. In this example program, whatever data is read on the monitored points is written to the control points. The direction of data flow to/from the 8051 is controlled by P3.2, (high = in). P3.3 latches new data to the process.

ADDRESS	MNEMONIC	COMMENT
	.org 0000h	
	.equ mon1,70h	;store first 8 monitored points
	.equ mon2,71h	;store second 8 monitored points
modezero:	clr p3.2	;load data from process to input
		;registers
	setb p3.2	;enable data shift in
	acall monit	;get first byte
	mov mon1,a	;store first byte
	acall monit	;get second byte
	mov mon2,a	;store second byte
	clr p3.2	;enable data to be shifted out

Continued

ADDRESS	MNEMONIC	COMMENT
Continued		

```
                acall conit     ;start sending data, second byte first
                mov a,monl      ;get first byte
                acall conit     ;send first byte
                clr p3.3        ;latch data to output latches
                setb p3.3       ;end latch strobe
                sjmp modezero   ;loop for any new input
;
;the routine that reads the monitored points follows
;
monit:          mov scon,#10h   ;set mode 0 and enable reception
                                ;reset RI
here:           jnb scon.0,here ;wait for end of reception
                mov scon,#00h   ;clear receive enable and interrupt bit
                mov a,sbuf      ;read byte received
                ret             ;return to calling program
;
;the routine that sends the control data follows
;
conit:          mov scon,#00h   ;set mode 0 and clear all interrupt bits
                mov sbuf,a      ;start transmission
wait:           jnb scon.1,wait ;wait until transmission complete
                ret             ;return to calling program
                .end
```

 COMMENT _____

Note that in both the transmit and receive cases the interrupt bit must go high before the subroutine can be ended.

The data transmit and reception time is so short that interrupt-driven schemes are not efficient.

Mode 1: Standard 8-Bit UART Mode

In Chapter 7, several simple communication programs are studied that use the serial port configured as mode 1, the standard UART mode normally used to communicate in 8-bit ASCII code. Only seven bits are needed to encode the entire set of ASCII characters. The eighth bit can be used for even or odd parity or ignored completely. Asynchronous data transmission requires a start and stop bit to enable the receiving circuitry to detect the start and finish of a complete character. A total of ten bits is needed to transmit the 7-bit ASCII character, as shown in Figure 9.5.

Transmission begins whenever data is written to SBUF. It is the responsibility of the programmer to ensure that any previous character has been transmitted by inspecting the TI bit in SCON for a set condition. Data transmission begins with a high-to-low start bit transition on TXD that signals receiving circuitry that a new character is about to arrive. The 8-bit character follows, LSB first and MSB parity bit last, and then the stop bit, which is high for one bit period. If another character follows immediately, a new start bit is signaled by a high-to-low transition; otherwise, the line remains high. The width of each transmitted data bit is controlled by the baud rate clock used. The receiver must use the same baud rate as the transmitter, or it reads the data at the wrong time in the character stream.

FIGURE 9.5 Asynchronous 8-Bit Character Used in Mode 1

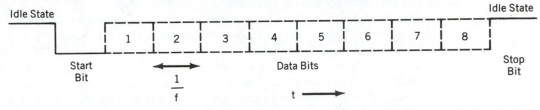

Reception begins if the REN bit is set in SCON and a high-to-low transition is sensed on RXD. Data bits are sampled at the baud rate in the center of the bit duration period. The received character is loaded into SBUF and the stop bit into SCON bit 2 (RB8) *if* the RI bit in SCON is cleared, indicating that the program has read the previous character; *and* either SM2 in SCON is cleared or SM2 is set and the received stop bit is high, which is the normal state for stop bits.

If these conditions are met, then SBUF is loaded with the received character, and RI is set. If the conditions are not met, the character is ignored, RI is not set, and the receive circuitry awaits the next start bit.

The restriction that a new character is not accepted unless RI is cleared seems reasonable. Data is lost if either the previous byte is overwritten or the new byte discarded, which is the action taken by the 8051. The restriction on SM2 and the stop bit are not as obvious. Normally, SM2 will be set to 0, and the character will be accepted no matter what the state of the stop bit. Software can check RB8 to ensure that the stop bit is correct before accepting the character if that is thought to be important.

Possible reasons for setting SM2 to force reception only when the stop bit is a 1 could be useful if the transmitter has the ability to change the stop bit from the normal high state. If the transmitter has this capability, then the stop bit can serve as an address bit in a multiprocessor environment where many loop microcontrollers are all receiving the same transmission. Only the microcontroller that has SM2 cleared can receive characters ending in either of the stop bit states. If all the microcontrollers but one have SM2 set, then all data transmissions ending in a low stop bit interrupt the unit with SM2 = 0; the rest ignore the data. Transmissions ending in a high stop bit can interrupt all microcontrollers.

Transmitters with the capability to alter the stop bit state are not standard. The 8051 communication modes 2 and 3 use the SM2 bit for multiprocessing. Mode 1 is not needed for this use.

In summary, mode 1 should be used with SM2 cleared, as a standard 8-bit UART, with software checks for proper stop bit magnitude if needed. As discussed in Chapter 7, the baud rate for modes 1 and 3 are determined by the overflow rate of timer 1, which is usually configured as an auto-reload timer. PCON bit 7 (SMOD) can double the baud rate when set.

Modeone

Mode 1 is most likely to be used in a dedicated system where the 8051 serial port is connected to a single similar port. A program that transmits and receives large blocks of data on an interrupt-driven basis is developed to investigate some problems common to data interchange programs.

To the main program, interrupt-driven communication routines are transparent: Data appears in RAM as it is received and disappears from RAM as it is transmitted. In both cases, the link between the main program and the interrupt-driven communication subroutines are areas of RAM called buffers. These buffers serve to store messages that are to be sent and messages that are received.

Each buffer area is defined by two memory pointers. One pointer contains the address of the top of the buffer, or the location in RAM where the next character is to be stored, and the second contains the address of the next character to be read. The buffers are named "inbuf," for use in storing characters as they are received, and "outbuf," for storing characters that are to be sent. The pointers to the tops of the buffers are named "intop" and "outop," respectively, while the pointers to the next character to be read are named "inplace" and "outplace."

The two buffers work in exactly the same way. The receive subroutine fills inbuf as characters are received and updates intop as it operates. The main program empties inbuf as it can and keeps inplace pointing to the next character to be read. The main program fills outbuf, while keeping outop updated to point to the next character to be stored. The transmission subroutine empties outbuf as it can and keeps outplace pointing to the next character to be read from outbuf.

These actions continue until the pointer to the top of the buffer equals the pointer to the next character. The buffer is now empty, and the pointers can be reset to the bottom of the buffer.

The buffer areas and pointers may be summarized as follows:

Outbuf: An area of RAM that holds characters to be transmitted

Outop: Pointer to outbuf that holds the address of the next character to be stored by the main program for transmission

Outplace: Pointer to outbuf that holds the address of the next character to be transmitted by the transmit subroutine

Inbuf: An area of RAM that holds received characters

Intop: Pointer to inbuf that holds the address of the next character received by the receive subroutine

Inplace: Pointer to inbuf that holds the address of the next character to be read by the main program

The main program and the transmit subroutine does not read data from a buffer whenever the place pointer equals the top pointer, which indicates that the buffer is empty.

The programmer has to make an estimate of how large the buffers need to be. Sometimes the general nature of the data is known when the system is in the design phase. The programmer(s) for the two computers that are communicating can define message length and frequency, arriving at a worst-case buffer size.

If the 8051 is part of a peripheral, such as a printer, that randomly receives large quantities of data, then the buffer size is fixed at an economic and competitive number using external RAM. For short and infrequent messages, internal RAM may suffice. In both cases, the receiving subroutine should have a means of communicating to the source of data when inbuf is becoming full so that the data flow can be suspended while inbuf is emptied. Our example program falls somewhere between these extremes; some external RAM will be needed, but not 32 kilobytes.

Registers R0 and R1 of register banks 0 and 1 are used effectively as pointers to the first 256d bytes of external RAM using MOVX instructions. For this example, the buffer sizes are fixed at 128d bytes each, although there is no need for them to be of equal size. Larger buffers can be constructed using the DPTR.

A program named "Modeone" handles communications between the 8051 and another computer using serial data mode 1. Two 128d byte buffers in external RAM store characters to be transmitted or received. R0 and R1 of register bank 0 keep track of data flow for the receive buffer inbuf, located in external RAM addresses 00h to 7Fh. R0 and R1 of register bank 1 serve the transmit buffer outbuf, external RAM addresses 80h to FFh. R0 is the place pointer, R1 the top pointer to the buffers. The baud rate is set by timer 1 in the auto-reload mode to 1200 bits per second. Port pin 3.2 is set high when inbuf is 1 byte from a full condition.

ADDRESS	MNEMONIC	COMMENT
	.org 0000h	
modeone:	sjmp over	;jump over serial interrupt address
	.org 0023h	;serial interrupt vectors to this ;address
	push psw	;save register bank status
	push acc	;save A
	jbc scon.0,rcve	;serve the received data first
	jbc scon.1,xmit	;transmit data as a second priority
	sjmp go	;should never get to this jump
rcve:	clr psw.3	;select register bank 0 pointers to ;inbuf
	mov a,sbuf	;get received character
	movx @r1,a	;store character at top of inbuf
	inc r1	;increment top address of inbuf
	cjne r1,#7eh,rok	;see if inbuf is almost full
	setb p3.2	;signal data source of full condition
	sjmp full	
rok:	clr p3.2	;remove full signal to source
full:	jbc scon.1,xmit	;see if transmit interrupt also ;occurred
	sjmp go	;if not then return
xmit:	setb psw.3	;select bank 1 pointers to outbuf
	mov a,r0	;compare R0 and R1 for equality
	cjne a,09h,mor	;internal RAM address 09h = R1, ;bank one
	mov r0,#80h	;reset both pointers to bottom of ;outbuf
	mov r1,r0	
	sjmp go	;buffer is empty; return
mor:	movx a,@r0	;get next character to be transmitted
	mov sbuf,a	;begin transmission
	inc r0	;point to next transmit character
go:	pop acc	;restore A and PSW, return
	pop psw	
	reti	

```
;
;the main program begins here; for the purpose of this example, the
;main program will send the character T repeatedly
;
```

over:	mov sp,#10h	;set SP above register bank one
	mov r0,#00h	;set inbuf pointers to bottom of

Continued

ADDRESS	MNEMONIC	COMMENT
Continued		
		;buffer
	mov r1,#00h	
	setb psw.3	;set outbuf pointers to bottom of ;buffer
	mov r0,#80h	
	mov r1,#80h	
	clr p3.2	;signal source that inbuf is not full
	mov scon,#50h	;set serial port mode 1; enable ;receiver
	mov th1,#0bbh	;set TH1 for 187d (baud rate = 1208)
	orl pcon,#80h	;set SMOD to double baud rate
	mov tmod,#20h	;set timer 1 to auto-reload mode
	mov tcon,#40h	;start T1 to generate baud clock
	mov ie,#90h	;enable only serial interrupt
loop:	clr psw.3	;set pointers to inbuf and get ;any data
	mov a,r0	;inbuf is empty when intop = inplace
	cjne a,01h,rd	;R1 of bank 0 is direct address 01h
	mov r0,#00h	;buffer empty, reset pointers to ;bottom
	mov r1,#00h	
	sjmp send	;send the character T
rd:	movx a,@r0	;get character
	inc r0	;point to next character
send:	setb psw.3	;set pointers to outbuf and store data
	cjne r1,#00h,sd	;see if outbuf is full and loop if so
	sjmp loop	;R1 rolls over from FFh to 00h if full
sd:	mov a,#'T'	;put the character T in outbuf
	movx @r1,a	;store a T in outbuf
	inc r1	;point to top of outbuf
	cjne r1,#81h,loop	;initiate transmission if first T
	setb scon.1	;start transmission process for first ;T
	sjmp loop	;continue
	.end	

────▷ COMMENT ──────────────────────────────────────

Note that the program has to initiate the first interrupt for the first character that is stored in a previously empty outbuf. If the first interrupt action were not done, transmission would never take place, as the TI bit would remain a 0. The 0 state of the TI bit is ambiguous: It can mean that the transmitter is busy sending a byte or that no activity is taking place at all. The 1 state of TI is specific: A byte has been transmitted, and SBUF can receive the next byte.

The example program fills outbuf quickly, until outop rolls over to 00h. Outbuf is emptied until outplace rolls over also, and outbuf is re-initialized to 80h. Received data is always read before inbuf can fill up, as there is very little for the program to do. Adding a time delay in the program ensures that inbuf grows beyond one byte.

Continued

COMMENT

The data source should cease sending data to the 8051 until port 3.2 goes low. In this example, "full" is arbitrarily set at one byte below the maximum capacity of inbuf. The actual number for a full condition should be set at maximum capacity less the response time of the source expressed in characters.

No feedback from the source to the 8051 has been provided for halting transmission of data from the 8051. Feedback can be accomplished by using one of the $\overline{INT}$ lines as an input from the source to signal a full condition.

Modes 2 and 3: Multiprocessor

Modes 2 and 3 are identical except for the baud rate. Mode 2 uses a baud rate of f/32 if SMOD (PCON.7) is cleared or f/64 if SMOD is set. For our 16 megahertz example, this results in baud rates of 500000 and 250000 bits per second, respectively. Pulse rates of these frequencies require care in the selection and installation of the transmission lines used to carry the data.

Baud rates for mode 3 are programmable using the overflows of timer 1 exactly as for data mode 1. Baud rates as high as 83333 bits per second are possible using a 16 megahertz crystal. These rates are compatible with RS 485 twisted-pair transmission lines.

Data transmission using modes 2 and 3 features eleven bits per character, as shown in Figure 9.6. A character begins with a start bit, which is a high-to-low transition that lasts one bit period, followed by 8 data bits, LSB first. The tenth bit of this character is a programmable bit that is followed by a stop bit. The stop bit remains in a high state for a minimum of one bit period.

Inspection of Figures 9.5 and 9.6 reveals that the only difference between mode 1 and mode 2 and 3 data transmission is the addition of the programmable tenth bit in mode 2 and 3.

When the 8051 transmits a character in mode 2 and 3, the eight data bits are whatever value is loaded in SBUF. The tenth bit is the value of bit SCON.3, named TD8. This bit can be cleared or set by the program. Interrupt bit TI (SCON.1) is set after a character has been transmitted and must be reset by program action.

Characters received using mode 2 and 3 have the eight data bits placed in SBUF and the tenth bit is in SCON.2, called RB8, if certain conditions are true. Two conditions apply to receive a character. First, interrupt bit RI (SCON.0) must be cleared before the last bit of the character is received, and second, bit SM2 (SCON.5) must be a 0 or the tenth bit must be a 1. If these conditions are met, then the eight data bits are loaded in SBUF, the tenth bit is placed in RB8, and the receive interrupt bit RI is set. If these conditions are not met, the character is ignored, and the receiving circuitry awaits the next start bit.

The significant condition is the second. If RI is set, then the software has not read the previous data (or forgot to reset RI), and it would serve no purpose to overwrite the

FIGURE 9.6   Asynchronous 9-Bit Character Used in Modes 2 and 3

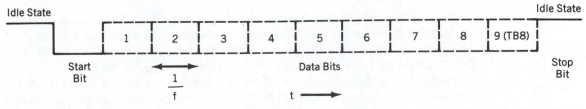

data. Clearing SM2 to 0 allows the reception of multiprocessor characters transmitted in mode 2 and 3. Setting SM2 to 1 prevents the reception of those characters that have bit ten equal to 0. Put another way, if bit ten is a 1, then reception always takes place; SM2 is ignored. *If* bit ten is a 0 then *only* those receivers with SM2 set to 0 are interrupted.

Mode 2 and 3 has been included in the 8051 specifically to enhance the use of multiple 8051s that are connected to a common loop in a multiprocessor configuration. The term *multiprocessing* implies many processors acting in some unified manner and connected so that data can be interchanged between them. When the processors are connected in a loop configuration, then there is generally a controlling or "talker" processor that directs the activities of the remainder of the loop units, or "listeners."

One particular characteristic of a talker–listener loop is the frequent transmission of data between the talker and individual listeners. All data broadcast by the talker is received by all the listeners, although often the data is intended only for one or a few listeners. At times, data is broadcast that is meant to be used by all the listeners.

There are many ways to handle the addressing problem. Systems that use standard UART technology, such as mode 1, can assign unique addresses to all the listeners. Each message from the talker can begin with the address of the particular listener for which it is intended. When a message is sent, all the listeners process the message and react only if the address that begins the message matches their assigned addresses. If messages are sent frequently, the listeners will waste a lot of processing time rejecting those messages not addressed to them.

Mode 2 and 3 reduces processing time by enabling character reception based upon the state of SM2 in a listener and the state of bit ten in the transmitted character. A single strategy is used to enable a few listeners to receive data while the majority ignore the transmissions.

All listeners initially have SM2 set to 0, the normal reset state, and receive all multiprocessor messages. Each listener has a unique address. The talker addresses each of the listeners that are not of interest and commands them to set SM2 to 1, leaving the listeners to which communication is desired with SM2 cleared to 0. All characters from the talker to the unique listeners are then sent with bit ten set to 0. The listener(s) with SM2 cleared receive the data; those with SM2 set ignore the data due to the condition of bit ten. Communication with all listeners is done by setting bit ten to 1, which enables reception of characters with no regard as to the state of SM2.

A variation of this strategy is to have all listeners set SM2 to 1 upon power-up. All address messages have a 1 in bit position ten, so all listeners receive and process any address message to see whether action is required. Listeners chosen are commanded in the address message to set SM2 to 0, and data communication proceeds with bit ten cleared to 0.

The multiprocessing strategy works best when there is extensive data interchange between the talker and each individual listener. Frequent changes of listeners with little data flow results in heavy address usage and subsequent interruption of all listeners to process the address messages.

Modethree

A multiprocessor configuration that demonstrates the use of mode 3 is shown in Figure 9.7. An RS 485 twisted-pair transmission line is used to form a loop that has 15d 8051 microprocessors connected to the lines so that all data on the loop is common to all serial ports. The 8051 has been programmed to be the talker, and the rest are listeners.

The purpose of the loop is to collect ten data bytes from each listener, in sequential order. All listeners initialize SM2 to 1 after power-up, and the talker configures all address

FIGURE 9.7 Communication Loop Used for Modethree Program

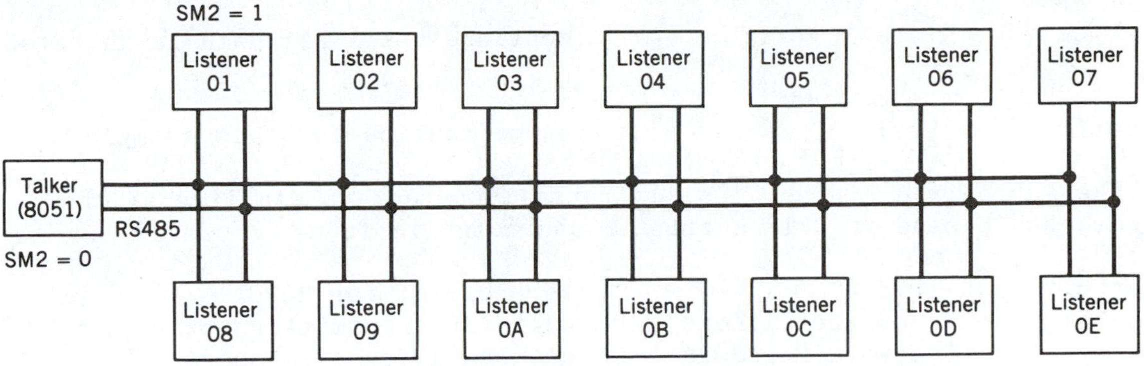

messages using a 1 in bit ten. Addressed listeners transmit ten data characters to the talker with bit ten set to 0. The talker has SM2 set to 0 so that all communications from listeners are acknowledged. Data characters from a listener to the talker are ignored by the remaining listeners. At the end of the ten data bytes, the addressed listener resets SM2 to 1. The data rate is set by timer 1 in the auto-reload mode to be 83333 baud. That portion of the talker and listener program that has to do with setting up the multiprocessor environment will be programmed.

The messages that are sent from the talker to the listeners are called "canned" because the contents of each is known when the program is written; the messages can be placed in ROM for later use. The subroutine "sendit" in the talker program can send canned messages of arbitrary length, as long as each message ends in the character $.

Message contents from the listeners to the talker are not known when the program is written. A version of sendit, "sndat," can still be used if the message is constructed in the same manner as the canned messages in the ROM of the talker program.

The program "Modethree" sends a canned address message to each of Fh listeners on a party-line loop using serial data mode 3. All canned messages are transmitted with bit ten set to 1; all received data from the addressed listener has bit ten set to 0. SM2 is set in all listeners and reset in the talker.

ADDRESS	MNEMONIC	COMMENT
	.org 0000h	
modethree:	mov scon,#0dah	;set mode 3, REN, TB8 and TI, ;clear SM2
	mov th1,#0ffh	;set TH1 for 83333d overflow rate
	orl pcon,#80h	;set SMOD
	mov tmod,#20h	;set timer 1 to auto-reload mode
	mov tcon,#40h	;start T1 to generate baud clock
	mov dptr,#add1	;send first listener message
talker:	acall sendit	
	acall getit	;"getit" is a data reception ;routine
	mov dptr,#add2	;send second listener message
	acall sendit	

Continued

ADDRESS	MNEMONIC	COMMENT

Continued

```
                acall getit         ;continue until all data is gathered
                ; .....
                sjmp over
getit:          ret                 ;dummy routine for this example
;
;the subroutine "sendit" will transmit characters starting at the
;address passed in DPTR until a $ character is found
;
sendit:         clr a               ;zero offset for MOVC
here:           jnb scon.1,here     ;wait for transmitter not busy
                movc a,@a+dptr      ;get character of message
                mov sbuf,a          ;send character
                cjne a,#'$',out     ;if a $ then return to calling
                                    ;program
                ret
out:            inc dptr            ;point to next character
                sjmp sendit         ;continue until done
;
;the canned address messages are assembled in ROM next
;
add1:           .db "01$"           ;address message for listener 1
add2:           .db "02$"           ;address message for listener 2
; ............                      ;continue for all listeners
add15:          .db "0f$"
                .end
;the program "listener" recognizes its address and responds with
;10 data characters; the data message is built in RAM, and ends
;with a $ character; for this example, the data is gotten by reading
;port 1 ten times and storing the data; this is the program for
;listener 01
;
listener:       org 0000h
                mov scon,#0f2h      ;set mode3, SM2, REN, TI; clear TB8,
                                    ;RI, RB8
                mov th1,#0ffh       ;set TH1 for 83333d overflow rate
                orl pcon,#80h       ;set SMOD to double baud rate
                mov tmod,#20h       ;set timer 1 to auto-reload mode
                mov tcon,#40h       ;start T1 to generate baud clock
who:            jnb scon.0,who      ;look for the first address
                                    ;character
                clr scon.0          ;first character, clear receive flag
                mov a,sbuf          ;get character
                cjne a,#'0',no      ;compare against expected address
nxt:            jnb scon.0,nxt      ;first character correct, get second
                clr scon.0          ;second character, clear receive
                                    ;flag
```

Continued

ADDRESS	MNEMONIC	COMMENT
	mov a,sbuf	;check next character
	cjne a,#'l',no	
ok:	jnb scon.0,ok	;wait for $ and then send data
	clr scon.0	
	cjne a,#'$',no	;if not $ then reset
	sjmp sendata	
no:	jnb scon.0,no	;wait for $ and then loop
	clr scon.0	
	mov a,sbuf	;get character
	cjne a,#'$',no	;loop until $ found
	sjmp who	;loop until proper address sent
sendata:	mov r0,#50h	;build the message in RAM starting ;at 50h
	mov rl,#0ah	;set Rl to count data bytes from ;port 1
indat:	mov @r0,pl	;get data from port 1 to RAM
	inc r0	;point to next RAM location
	djnz rl,indat	;continue until 10d bytes are stored
	mov @r0,#'$'	;finish data string with a ;$ character
	mov r0,#50h	;reset R0 to point to start of ;message
sndat:	jnb scon.1,sndat	;wait for transmitter empty
	mov a,@r0	;get character from message
	mov sbuf,a	;load SBUF for transmission
	inc r0	;point to next character
	cjne a,#'$',sndat	;look for $ then stop transmission
	sjmp listener	;loop for next cycle
	.end	

 COMMENT

The inclusion of the $ character in each message is useful both as a check for the end of a message and to reset a listener that somehow misses one of the three characters expected in an address. If a listener misses a character, due to noise for example, it will get to the "no" label within one or two characters. The next $ will reset the listener program back to the "who" label.

Programs that interchange data must be written to eliminate any chance of a receiving unit getting caught in a trap waiting for a predetermined number of characters. Common schemes that accomplish this goal use special "end-of-message" characters, as in the case of Modethree, or set timers to interrupt the receiving program if the data is not received within a certain period of time.

Much more elaborate protocols than those used here in this example would be used by the listeners when sending data to the talker. There is always the possibility that errors will occur due to noise or the improper operation of another listener interfering. The talker may store these errors. Error-checking bytes may be added to the data stream so that the talker can verify that the string of characters is error free.

Summary

Four serial data communication modes for the 8051 are covered in this chapter:

Mode 0: High-speed, 8-bit shift register; one baud rate of f/12

Mode 1: Standard 8-bit UART; variable baud rate using timer 1 overflows

Mode 2: Multiprocessor 9-bit UART; two baud rates of f/32 and f/64

Mode 3: Multiprocessor 9-bit UART; variable baud rate using timer 1 overflows

Programs in this chapter use these modes and feature several standard communication techniques:

High-speed shift register data gathering

Interrupt-driven transmit and receive buffers

Sending preprogrammed, or canned, messages

Problems

1. Explain why mode 0 is not suitable for 8051 communications.

2. How much clock skew, in terms of clock period, can transmitted data using mode 0 have before data is shifted in error?

3. Repeat Problem 2 for data reception.

4. Assume you are determined to use mode 0 as a communication mode from one 8051 to another. Outline a system of hardware and software that would allow this. Hint: A "buffer" is needed.

5. Sketch the mode 1 no parity ASCII serial characters U, 0, and w.

6. Many communication terminals can determine the baud rate of standard (mode 1) characters by making measurements on the first few "fill" characters received. Outline a program strategy that would set the 8051 baud rate automatically based upon the first character received.

7. Character transmission can be done by using a time delay greater than the character time before moving a new byte to SBUF. Explain why character reception must use an interrupt flag if all characters are to be received.

8. ASCII characters can have even (number of ones), odd, or no parity using bit 7 as a parity bit. Write a program that checks the incoming data for odd parity and sets a flag if the parity is incorrect.

9. Write a program that converts odd parity bytes to even parity bytes (bit 7 is the parity bit).

10. An overrun is said to occur in data reception whenever a new byte of data is received before the previously received byte has been read. Discuss two methods by which overruns might be detected by the 8051 program.

11. List two reasons why stop bits are used in asynchronous communications.

12. A framing error is said to have occurred if the stop bit is not a logic high. What mode(s) can detect a framing error?

13. Why is it necessary for the main program (see "Modeone") to set the TI bit to begin the transmission of a string of characters using interrupt-driven routines? Name another way for the main program to initiate transmission.

14. Determine if an 8051 in mode 1 can communicate with an 8051 in mode 3.

15. Modify the "Modeone" program to use 4K byte buffers.

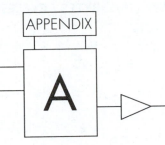

A

8051 Operational Code Mnemonics

Appendix A lists two arrangements of mnemonics for the 8051: by function, and alphabetically. The mnemonic definitions differ from that of the original manufacturer (Intel Corporation) by the names used for addresses or data; for example, "add" is used to represent an address in internal RAM, while Intel uses the name "direct." The author believes that the names used are clearer than those used by Intel. Appendix A also includes an alphabetical listing of the mnemonics using Intel names. There is **no** difference between the mnemonics when real numbers replace the names. For example; MOV add,#n and MOV direct,#data become MOV 10h,#40h when the number 10h replaces the internal RAM address (add/direct), and 40h replaces the number (#n/# data).

Mnemonics, Arranged by Function

Arithmetic

MNEMONIC	DESCRIPTION	BYTES	CYCLES	FLAGS
ADD A,Rr	A+Rr → A	1	1	C OV AC
ADD A,add	A+(add) → A	2	1	C OV AC
ADD A,@Rp	A+(Rp) → A	1	1	C OV AC
ADD A,#n	A+n → A	2	1	C OV AC
ADDC A,Rr	A+Rr+C → A	1	1	C OV AC
ADDC A,add	A+(add)+C → A	2	1	C OV AC
ADDC A,@Rp	A+(Rp)+C → A	1	1	C OV AC
ADDC A,#n	A+n+C → A	2	1	C OV AC
DA A	Abin → Adec	1	1	C
DEC A	A−1 → A	1	1	
DEC Rr	Rr−1 → Rr	1	1	
DEC add	(add)−1 → (add)	2	1	
DEC @Rp	(Rp)−1 → (Rp)	1	1	

Continued

Arithmetic

MNEMONIC	DESCRIPTION	BYTES	CYCLES	FLAGS
				Continued
DIV AB	A/B → AB	1	4	0 0V
INC A	A+1 → A	1	1	
INC Rr	Rr+1 → Rr	1	1	
INC add	(add)+1 → (add)	2	1	
INC @Rp	(Rp)+1 → (Rp)	1	1	
INC DPTR	DPTR+1 → DPTR	1	2	
MUL AB	A×B → AB	1	4	0 0V
SUBB A,Rr	A−Rr−C → A	1	1	C OV AC
SUBB A,add	A−(add)−C → A	2	1	C OV AC
SUBB A,@Rp	A−(Rp)−C → A	1	1	C OV AC
SUBB A,#n	A−n−C → A	2	1	C OV AC

Logic

MNEMONIC	DESCRIPTION	BYTES	CYCLES	FLAGS
ANL A,Rr	A AND Rr → A	1	1	
ANL A,add	A AND (add) → A	2	1	
ANL A,@Rp	A AND (Rp) → A	1	1	
ANL A,#n	A AND n → A	2	1	
ANL add,A	(add) AND A → (add)	2	1	
ANL add,#n	(add) AND n → (add)	3	2	
ORL A,Rr	A OR Rr → A	1	1	
ORL A,add	A OR (add) → A	2	1	
ORL A,@Rp	A OR (Rp) → A	1	1	
ORL A,#n	A OR n → A	2	1	
ORL add,A	(add) OR A → (add)	2	1	
ORL add,#n	(add) OR n → (add)	3	2	
XRL A,Rr	A XOR Rr → A	1	1	
XRL A,add	A XOR (add) → A	2	1	
XRL A,@Rp	A XOR (Rp) → A	1	1	
XRL A,#n	A XOR n → A	2	1	
XRL add,A	(add) XOR A → (add)	2	1	
XRL add,#n	(add) XOR n → (add)	3	2	
CLR A	00 → A	1	1	
CPL A	$\overline{A}$ → A	1	1	
NOP	PC+1 → PC	1	1	
RL A	A0←A7←A6..←A1←A0	1	1	
RLC A	C←A7←A6..←A0←C	1	1	C
RR A	A0→A →A6..→A1→A0	1	1	
RRC A	C→A7→A6..→A0→C	1	1	C
SWAP A	Alsn ↔ Amsn	1	1	

Data Moves

MNEMONIC	DESCRIPTION	BYTES	CYCLES	FLAGS
MOV A,Rr	Rr → A	1	1	
MOV A,add	(add) → A	2	1	
MOV A,@Rp	(Rp) → A	1	1	

Continued

MNEMONIC	DESCRIPTION	BYTES	CYCLES	FLAGS
MOV A,#n	n → A	2	1	
MOV Rr,A	A → Rr	1	1	
MOV Rr,add	(add) → Rr	2	2	
MOV Rr,#n	n → Rr	2	1	
MOV add,A	A → (add)	2	1	
MOV add,Rr	Rr → (add)	2	2	
MOV add1,add2	(add2) → (add1)	3	2	
MOV add,@Rp	(Rp) → (add)	2	2	
MOV add,#n	n → (add)	3	2	
MOV @Rp,A	A → (Rp)	1	1	
MOV @Rp,add	(add) → (Rp)	2	2	
MOV @Rp,#n	n → (Rp)	2	1	
MOV DPTR,#nn	nn → DPTR	3	2	
MOVC A,@A+DPTR	(A+DPTR) → A	1	2	
MOVC A,@A+PC	(A+PC) → A	1	2	
MOVX A,@DPTR	(DPTR)^ → A	1	2	
MOVX A,@Rp	(Rp)^ → A	1	2	
MOVX @Rp,A	A → (Rp)^	1	2	
MOVX @DPTR,A	A → (DPTR)^	1	2	
POP add	(SP) → (add)	2	2	
PUSH add	(add) → (SP)	2	2	
XCH A,Rr	A ↔ Rr	1	1	
XCH A,add	A ↔ (add)	2	1	
XCH A,@Rp	A ↔ (Rp)	1	1	
XCHD A,@Rp	Alsn ↔ (Rp)lsn	1	1	

Calls and Jumps

MNEMONIC	DESCRIPTION	BYTES	CYCLES	FLAGS
ACALL sadd	PC+2 → (SP); sadd → PC	2	2	
CJNE A,add,radd	[A<>(add)]: PC+3+radd → PC	3	2	C
CJNE A,#n,radd	[A<>n]: PC+3+radd → PC	3	2	C
CJNE Rr,#n,radd	[Rr<>n]: PC+3+radd → PC	3	2	C
CJNE @Rp,#n,radd	[(Rp)<>n]: PC+3+radd → PC	3	2	C
DJNZ Rr,radd	[Rr−1<>00]: PC+2+radd → PC	2	2	
DJNZ add,radd	[(add)−1<>00]: PC+3+radd → PC	3	2	
LCALL ladd	PC+3 → (SP); ladd → PC	3	2	
AJMP sadd	sadd → PC	2	2	
LJMP ladd	ladd → PC	3	2	
SJMP radd	PC+2+radd → PC	2	2	
JMP @A+DPTR	DPTR+A → PC	1	2	
JC radd	[C=1]: PC+2+radd → PC	2	2	
JNC radd	[C=0]: PC+2+radd → PC	2	2	
JB b,radd	[b=1]: PC+3+radd → PC	3	2	
JNB b,radd	[b=0]: PC+3+radd → PC	3	2	
JBC b,radd	[b=1]: PC+3+radd → PC; 0 → b	3	2	
JZ radd	[A=00]: PC+2+radd → PC	2	2	
JNZ radd	[A>00]: PC+2+radd → PC	2	2	
RET	(SP) → PC	1	2	
RETI	(SP) → PC; EI	1	2	

Boolean

MNEMONIC	DESCRIPTION	BYTES	CYCLES	FLAGS
ANL C,b	C AND b → C	2	2	C
ANL C,$\overline{b}$	C AND $\overline{b}$ → C	2	2	C
CLR C	0 → C	1	1	0
CLR b	0 → b	2	1	
CPL C	$\overline{C}$ → C	1	1	C
CPL b	$\overline{b}$ → b	2	1	
ORL C,b	C OR b → C	2	2	C
ORL C,$\overline{b}$	C OR $\overline{b}$ → C	2	2	C
MOV C,b	b → C	2	1	C
MOV b,C	C → b	2	2	
SETB C	1 → C	1	1	1
SETB b	1 → b	2	1	

Mnemonics, Arranged Alphabetically

MNEMONIC	DESCRIPTION	BYTES	CYCLES	FLAGS
ACALL sadd	PC+2 → (SP); sadd → PC	2	2	
ADD A,add	A+(add) → A	2	1	C OV AC
ADD A,@Rp	A+(Rp) → A	1	1	C OV AC
ADD A,#n	A+n → A	2	1	C OV AC
ADD A,Rr	A+Rr → A	1	1	C OV AC
ADDC A,add	A+(add)+C → A	2	1	C OV AC
ADDC A,@Rp	A+(Rp)+C → A	1	1	C OV AC
ADDC A,#n	A+n+C → A	2	1	C OV AC
ADDC A,Rr	A+Rr+C → A	1	1	C OV AC
AJMP sadd	sadd → PC	2	2	
ANL A,add	A AND (add) → A	2	1	
ANL A,@Rp	A AND (Rp) → A	1	1	
ANL A,#n	A AND n → A	2	1	
ANL A,Rr	A AND Rr → A	1	1	
ANL add,A	(add) AND A → (add)	2	1	
ANL add,#n	(add) AND n → (add)	3	2	
ANL C,b	C AND b → C	2	2	C
ANL C,$\overline{b}$	C AND $\overline{b}$ → C	2	2	C
CJNE A,add,radd	[A<>(add)]: PC+3+radd → PC	3	2	C
CJNE A,#n,radd	[A<>n]: PC+3+radd → PC	3	2	C
CJNE @Rp,#n,radd	[(Rp)<>n]: PC+3+radd → PC	3	2	C
CJNE Rr,#n,radd	[Rr<>n]: PC+3+radd → PC	3	2	C
CLR A	0 → A	1	1	
CLR b	0 → b	2	1	
CLR C	0 → C	1	1	0
CPL A	$\overline{A}$ → A	1	1	
CPL b	$\overline{b}$ → b	2	1	
CPL C	$\overline{C}$ → C	1	1	C
DA A	Abin → Adec	1	1	C
DEC A	A−1 → A	1	1	
DEC add	(add)−1 → (add)	2	1	
DEC @Rp	(Rp)−1 → (Rp)	1	1	
DEC Rr	Rr−1 → Rr	1	1	

Continued

MNEMONIC	DESCRIPTION	BYTES	CYCLES	FLAGS
DIV AB	A/B → AB	1	4	0 OV
DJNZ add,radd	[(add)−1<>00]: PC+3+radd → PC	3	2	
DJNZ Rr,radd	[Rr−1<>00]: PC+2+radd → PC	2	2	
INC A	A+1 → A	1	1	
INC add	(add)+1 → (add)	2	1	
INC DPTR	DPTR+1 → DPTR	1	2	
INC @Rp	(Rp)+1 → (Rp)	1	1	
INC Rr	Rr+1 → Rr	1	1	
JB b,radd	[b=1]: PC+3+radd → PC	3	2	
JBC b,radd	[b=1]: PC+3+radd → PC; 0 → b	3	2	
JC radd	[C=1]: PC+2+radd → PC	2	2	
JMP @A+DPTR	DPTR+A → PC	1	2	
JNB b,radd	[b=0]: PC+3+radd → PC	3	2	
JNC radd	[C=0]: PC+2+radd → PC	2	2	
JNZ radd	[A>00]: PC+2+radd → PC	2	2	
JZ radd	[A=00]: PC+2+radd → PC	2	2	
LCALL ladd	PC+3 → (SP); ladd → PC	3	2	
LJMP ladd	ladd → PC	3	2	
MOV A,add	(add) → A	2	1	
MOV A,@Rp	(Rp) → A	1	1	
MOV A,#n	n → A	2	1	
MOV A,Rr	Rr → A	1	1	
MOV add,A	A → (add)	2	1	
MOV add1,add2	(add2) → (add1)	3	2	
MOV add,@Rp	(Rp) → (add)	2	2	
MOV add,#n	n → (add)	3	2	
MOV add,Rr	Rr → (add)	2	2	
MOV b,C	C → b	2	2	
MOV C,b	b → C	2	1	C
MOV @Rp,A	A → (Rp)	1	1	
MOV @Rp,add	(add) → (Rp)	2	2	
MOV @Rp,#n	n → (Rp)	2	1	
MOV DPTR,#nn	nn → DPTR	3	2	
MOV Rr,A	A → Rr	1	1	
MOV Rr,add	(add) → Rr	2	2	
MOV Rr,#n	n → Rr	2	1	
MOVC A,@A+DPTR	(A+DPTR) → A	1	2	
MOVC A,@A+PC	(A+PC) → A	1	2	
MOVX A,@DPTR	(DPTR)^ → A	1	2	
MOVX A,@Rp	(Rp)^ → A	1	2	
MOVX @DPTR,A	A → (DPTR)^	1	2	
MOVX @Rp,A	A → (Rp)^	1	2	
NOP	PC+1 → PC	1	1	
MUL AB	AxB → AB	1	4	0 OV
ORL A,add	A OR (add) → A	2	1	
ORL A,@Rp	A OR (Rp) → A	1	1	
ORL A,#n	A OR n → A	2	1	
ORL A,Rr	A OR Rr → A	1	1	
ORL add,A	(add) OR A → (add)	2	1	
ORL add,#n	(add) OR n → (add)	3	2	
ORL C,b	C OR b → C	2	2	C
ORL C,$\overline{b}$	C OR $\overline{b}$ → C	2	2	C

Continued

MNEMONIC	DESCRIPTION	BYTES	CYCLES	FLAGS
				Continued
POP add	(SP) → (add)	2	2	
PUSH add	(add) → (SP)	2	2	
RET	(SP) → PC	1	2	
RETI	(SP) → PC; EI	1	2	
RL A	A0←A7←A6. .←A1←A0	1	1	
RLC A	C←A7←A6. .←A0←C	1	1	C
RR A	A0→A7→A6. .→A1→A0	1	1	
RRC A	C→A7→A6. .→A0→C	1	1	C
SETB b	1 → b	2	1	
SETB C	1 → C	1	1	1
SJMP radd	PC+2+radd → PC	2	2	
SUBB A,add	A−(add)−C → A	2	1	C OV AC
SUBB A,@Rp	A−(Rp)−C → A	1	1	C OV AC
SUBB A,#n	A−n−C → A	2	1	C OV AC
SUBB A,Rr	A−Rr−C → A	1	1	C OV AC
SWAP A	Alsn ↔ Amsn	1	1	
XCH A,add	A ↔ (add)	2	1	
XCH A,@Rp	A ↔ (Rp)	1	1	
XCH A,Rr	A ↔ Rr	1	1	
XCHD A,@Rp	Alsn ↔ (Rp)lsn	1	1	
XRL A,add	A XOR (add) → A	2	1	
XRL A,@Rp	A XOR (Rp) → A	1	1	
XRL A,#n	A XOR n → A	2	1	
XRL A,Rr	A XOR Rr → A	1	1	
XRL add,A	(add) XOR A → (add)	2	1	
XRL add,#n	(add) XOR n → (add)	3	2	

MNEMONIC ACRONYMS

add	Address of the internal RAM from 00h to FFh.
ladd	Long address of 16 bits from 0000h to FFFFh.
radd	Relative address, a signed number from −128d to +127d.
sadd	Short address of 11 bits; complete address = PC11−PC15 and sadd.
b	Addressable bit in internal RAM or a SFR.
C	The carry flag.
lsn	Least significant nibble.
msn	Most significant nibble.
n	Any immediate 8 bit number from 00h to FFh.
Rr	Any of the eight registers, R0 to R7 in the selected bank.
Rp	Either of the pointing registers R0 or R1 in the selected bank.
[ ]:	IF the condition inside the brackets is *true,* THEN the action listed will occur; ELSE go to the next instruction.
∧	External memory location.
( )	Contents of the location inside the parentheses.

Note that flags affected by each instruction are shown where appropriate; any operations that affect the PSW address may also affect the flags.

Intel Corporation Mnemonics, Arranged Alphabetically

MNEMONIC	DESCRIPTION	BYTES	CYCLES	FLAGS
ACALL addr11	PC+2 → (SP); addr11 → PC	2	2	
ADD A,direct	A+(direct) → A	2	1	C OV AC
ADD A,@Ri	A+(Ri) → A	1	1	C OV AC
ADD A,#data	A+#data → A	2	1	C OV AC
ADD A,Rn	A+Rn → A	1	1	C OV AC
ADDC A,direct	A+(direct)+C → A	2	1	C OV AC
ADDC A,@Ri	A+(Ri)+C → A	1	1	C OV AC
ADDC A,#data	A+#data+C → A	2	1	C OV AC
ADDC A,Rn	A+Rn+C → A	1	1	C OV AC
AJMP addr11	addr11 → PC	2	2	
ANL A,direct	A AND (direct) → A	2	1	
ANL A,@Ri	A AND (Ri) → A	1	1	
ANL A,#data	A AND #data → A	2	1	
ANL A,Rn	A AND Rn → A	1	1	
ANL direct,A	(direct) AND A → (direct)	2	1	
ANL direct,#data	(direct) AND #data → (direct)	3	2	
ANL C,bit	C AND bit → C	2	2	C
ANL C,bit̄	C AND bit̄ → C	2	2	C
CJNE A,direct,rel	[A<>(direct)]: PC+3+rel → PC	3	2	C
CJNE A,#data,rel	[A<>n]: PC+3+rel → PC	3	2	C
CJNE @Ri,#data,rel	[(Ri)<>n]: PC+3+rel → PC	3	2	C
CJNE Rn,#data,rel	[Rn<>n]: PC+3+rel → PC	3	2	C
CLR A	0 → A	1	1	
CLR bit	0 → bit	2	1	
CLR C	0 → C	1	1	0
CPL A	$\overline{A}$ → A	1	1	
CPL bit	$\overline{bit}$ → bit	2	1	
CPL C	$\overline{C}$ → C	1	1	C
DA A	Abin → Adec	1	1	C
DEC A	A−1 → A	1	1	
DEC direct	(direct)−1 → (direct)	2	1	
DEC @Ri	(Ri)−1 → (Ri)	1	1	
DEC Rn	Rn−1 → Rn	1	1	
DIV AB	A/B → AB	1	4	0 OV
DJNZ direct,rel	[(direct)−1<>00]: PC+3+rel → PC	3	2	
DJNZ Rn,rel	[Rn−1<>00]: PC+2+rel → PC	2	2	
INC A	A+1 → A	1	1	
INC direct	(direct)+1 → (direct)	2	1	
INC DPTR	DPTR+1 → DPTR	1	2	
INC @Ri	(Ri)+1 → (Ri)	1	1	
INC Rn	Rn+1 → Rn	1	1	
JB bit,rel	[b=1]: PC+3+rel → PC	3	2	
JBC bit,rel	[b=1]: PC+3+rel → PC; 0 → bit	3	2	
JC rel	[C=1]: PC+2+rel → PC	2	2	
JMP @A+DPTR	DPTR+A → PC	1	2	
JNB bit,rel	[b=0]: PC+3+rel → PC	3	2	
JNC rel	[C=0]: PC+2+rel → PC	2	2	
JNZ rel	[A>00]: PC+2+rel → PC	2	2	
JZ rel	[A=00]: PC+2+rel → PC	2	2	

Continued

MNEMONIC	DESCRIPTION	BYTES	CYCLES	FLAGS
LCALL addr16	PC+3 → (SP); addr16 → PC	3	2	
LJMP addr16	addr16 → PC	3	2	
MOV A,direct	(direct) → A	2	1	
MOV A,@Ri	(Ri) → A	1	1	
MOV A,#data	#data → A	2	1	
MOV A,Rn	Rn → A	1	1	
MOV direct,A	A → (direct)	2	1	
MOV direct,direct	(direct) → (direct)	3	2	
MOV direct,@Ri	(Ri) → (direct)	2	2	
MOV direct,#data	#data → (direct)	3	2	
MOV direct,Rn	Rn → (direct)	2	2	
MOV bit,C	C → bit	2	2	
MOV C,bit	bit → C	2	1	C
MOV @Ri,A	A → (Ri)	1	1	
MOV @Ri,direct	(direct) → (Ri)	2	2	
MOV @Ri,#data	#data → (Ri)	2	1	
MOV DPTR,#data16	#data16 → DPTR	3	2	
MOV Rn,A	A → Rn	1	1	
MOV Rn,direct	(direct) → Rn	2	2	
MOV Rn,#data	#data → Rn	2	1	
MOVC A,@A+DPTR	(A+DPTR) → A	1	2	
MOVC A,@A+PC	(A+PC) → A	1	2	
MOVX A,@DPTR	(DPTR)^ → A	1	2	
MOVX A,@Ri	(Ri)^ → A	1	2	
MOVX @DPTR,A	A → (DPTR)^	1	2	
MOVX @Ri,A	A → (Ri)^	1	2	
NOP	PC+1 → PC	1	1	
MUL AB	A×B → AB	1	4	0 OV
ORL A,direct	A OR (direct) → A	2	1	
ORL A,@Ri	A OR (Ri) → A	1	1	
ORL A,#data	A OR #data → A	2	1	
ORL A;Rn	A OR Rn → A	1	1	
ORL direct,A	(direct) OR A → (direct)	2	1	
ORL direct,#data	(direct) OR #data → (direct)	3	2	
ORL C,bit	C OR bit → C	2	2	C
ORL C,$\overline{\text{bit}}$	C OR $\overline{\text{bit}}$ → C	2	2	C
POP direct	(SP) → (direct)	2	2	
PUSH direct	(direct) → (SP)	2	2	
RET	(SP) → PC	1	2	
RETI	(SP) → PC; EI	1	2	
RL A	A0←A7←A6..←A1←A0	1	1	
RLC A	C←A7←A6..←A0←C	1	1	C
RR A	A0→A7→A6..→A1→A0	1	1	
RRC A	C→A7→A6..→A0→C	1	1	C
SETB bit	1 → bit	2	1	
SETB C	1 → C	1	1	1
SJMP rel	PC+2+rel → PC	2	2	
SUBB A,direct	A−(direct)−C → A	2	1	C OV AC
SUBB A,@Ri	A−(Ri)−C → A	1	1	C OV AC
SUBB A,#data	A−#data−C → A	2	1	C OV AC
SUBB A,Rn	A−Rn−C → A	1	1	C OV AC

Continued

MNEMONIC	DESCRIPTION	BYTES	CYCLES	FLAGS
SWAP A	Alsn ↔ Amsn	1	1	
XCH A,direct	A ↔ (direct)	2	1	
XCH A,@Ri	A ↔ (Ri)	1	1	
XCH A,Rn	A ↔ Rn	1	1	
XCHD A,@Ri	Alsn ↔ (Ri)lsn	1	1	
XRL A,direct	A XOR (direct) → A	2	1	
XRL A,@Ri	A XOR (Ri) → A	1	1	
XRL A,#data	A XOR #data → A	2	1	
XRL A,Rn	A XOR Rn → A	1	1	
XRL direct,A	(direct) XOR A → (direct)	2	1	
XRL direct,#data	(direct) XOR #data → (direct)	3	2	

ACRONYMS

addr11	Page address of 11 bits, which is in the same 2K page as the address of the following instruction.
addr16	Address for any location in the 64K memory space.
bit	The address of a bit in the internal RAM bit address area or a bit in an SFR.
C	The carry flag.
#data	An 8-bit binary number from 00 to FFh.
#data16	A 16-bit binary number from 0000 to FFFFh.
direct	An internal RAM address or an SFR byte address.
lsn	Least significant nibble.
msn	Most significant nibble.
rel	Number that is added to the address of the next instruction to form an address +127d or −128d from the address of the next instruction.
Rn	Any of registers R0 to R7 of the current register bank.
@Ri	Indirect address using the contents of R0 or R1.
[ ]:	IF the condition inside the brackets is *true,* THEN the action listed will occur; ELSE go to the next instruction.
∧	EXTERNAL memory location.
( )	Contents of the location inside the parentheses.

Note that flags affected by each instruction are shown where appropriate; any operations which affect the PSW address may also affect the flags.

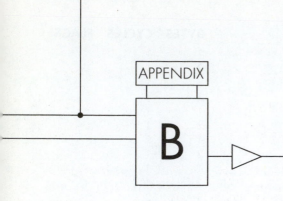

APPENDIX

B

How to Use the A51 Assembler

Introduction

In the early days of digital computing, (the 1940's), computers were programed in binary, resulting in programs that appeared as

```
1001001010101111
1111010111010110
0001011011110010
1100010101000101
.
.
.
```

and are generally unintelligible to anyone. Early in this process, programmers became tired of typing all those 1's and 0's, so a shorthand notation for binary (hexadecimal) was adopted to shorten the typing effort:

```
92AF
F5D6
16F2
C545
.
.
.
```

using 0–9 for binary 0000 to 1001, and A–F for binary 1010 to 1111. The result is still unintelligible, but more compact.

Each line of code is an instruction to the computer, and the programmers composed descriptions for the instructions that could be written as

```
Load the accumulator with a number
Move the accumulator to register 1
Move memory location 3 to location 2
```

The programmers translated these descriptions to the equivalent hex codes using pencil and paper. Very soon these long descriptions were shortened to

```
LODE A,NUM
MOVE 1,A
MOVE 2,3
   .
   .
   .
```

Mnemonics were born to speed up the programming process by retaining the essence of the instruction. Finally, programs became so long, and computing so inexpensive, programs that translated the mnemonics into their equivalent hex codes were written to facilitate the programming process.

The translation programs go by many names:

Interpreter: Translate each line of the program independently to an abbreviated ASCII (non-hex) equivalent

Assembler: Translate the entire program, as a whole, to hex

Compiler: Convert an Interpreter translation to hex

Then there are the "Cross" varieties, which assemble or compile code on computer A for use on computer B. This type of assembler is included with this book: A Cross—Assembler for the 8051, which runs on PC type computers. The assembler was written by David Akey of PseudoCode, Newport News, Virginia. The assembler is a file on the programming disk named A51.EXE.

Using the Assembler

The assembler included with this book is a student model that has been adapted from PseudoCodes' professional version. The student assembler has most of the capabilities of the professional version with these limitations:

No macro features

No options

The intent is to supply an assembler that is easy to use, enabling the student to get to the business of writing programs with a minimum of delay.

The Big Picture

An assembler is a translator machine. Computer programs, written using a defined set of rules (the syntax) are put into the assembler, and hex code pops out (if the syntax has been followed).

First, prepare a disk with the assembler program, A51.EXE. Next, place any input program you wish to have assembled on this disk.

The input program is in an ASCII text disk file that has been prepared by an editor program and that must obey these rules:

1. The file name has the extension .ASM (example: myfile.asm).
2. The file *must* be "pure" ASCII.

Many editor programs save text files using strange and potentially troublesome control characters. Save your text files in ASCII form.

The assembler produces two output files:

1. A file with the same name as the input ASCII text file, which has the extension .LST, is the assembled file complete with line numbers, memory addresses, hex codes, mnemonics, and comments. Any ERRORS found during assembly will be noted in the .LST file, at the point in the program where they occur.

2. A file with the same name as the input ASCII text file, which has the extension .OBJ, is the hex format file that can be loaded into the simulator and run.

Example:

A small program that blinks LEDs on an 8051 system is edited and saved as an ASCII file named try.asm.

```
              .org 4000h
      loop:   mov 90h,#0ffh      ;LEDs off
              acall time         ;delay
              mov 90h,#7fh       ;turn on LED one
              acall time
              mov 90h,#0bfh      ;turn on LED two
              acall time
              mov 90h,#3fh       ;both LEDs on
              acall time
              sjmp loop
      time:   mov r0,#03h
      in1:    mov r1,#00h
      in2:    mov r2,#00h
      wait:   djnz r2,wait
              djnz r1,in2
              djnz r0,in1
              ret
              .end
```

The .LST file, which is produced by the assembler, has these features:

Line	Address	Hex	Label	Mnemonics	Comments
000001	4000			.org 4000h	
000002	4000	7590FF	loop:	mov 90h,#0ffh	;LEDs off
000003	4003	1116		acall time	;delay
000004	4005	75907F		mov 90h,#7fh	;turn on LED one Line
000005	4008	1116		acall time	
000006	400A	7590BF		mov 90h,#0bfh	;turn on LED two
000007	400D	1116		acall time	
000008	400F	75903F		mov 90h,#3fh	;both LEDs on
000009	4012	1116		acall time	
000010	4014	80EA		sjmp loop	
000011	4016	7803	time:	mov r0,#03h	

Continued

Line	Address	Hex	Label	Mnemonics	Comments
000012	4018	7900	in1:	mov r1,#00h	
000013	401A	7A00	in2:	mov r2,#00h	
000014	401C	DAFE	wait:	djnz r2,wait	
000015	401E	D9FA		djnz r1,in2	
000016	4020	D8F6		djnz r0,in1	
000017	4022	22		ret	
000018	4023			.end	

The .OBJ file contains the hex code from the .LST file, together with special leading (:xxxxxxxx) and trailing characters (the last byte in each line which is a checksum) that can be loaded into the simulator or an EPROM burner:

```
:104000007590FF111675907F11167590BF1116757A
:10401000903F111680EA780379007A00DAFED9FA27
:03402000D8F622AD
:00000001FF
```

How to Assemble

After you have written your program using the mnemonics from Appendix A and saved the program in an ASCII text file, type:

```
A51 -s yourfile      (Note: No .ASM)
```

Where yourfile is the name of your ASCII program file. The -s prevents the assembler from including the symbol table at the end of your program. For the example program, we type: a51 -s try. The result is TRY.LST and TRY.OBJ

The assembler will assemble your program and inform you of any errors that are found. You can type the .LST file to the computer screen or print the listing to a printer. All errors *in syntax* will be shown by the assembler in the .LST file. Keep in mind that a program that has been successfully assembled is not guaranteed to work; it is only grammatically correct. (One can write sentences in English that are grammatically correct but make no sense, such as "see any government form"). Re-edit your program until assembly is successful.

Assembler Directives

An assembler is a program and has instructions just as any program. These are called "directives" or "pseudo operations" because they inform the assembler what to do with the mnemonics that it is to assemble. The pseudo ops are distinctly different from the mnemonics of the computer code being assembled so that they stand out in the program listing. For the PseudoCode assembler, they are

```
.org xxxx    ORiGinate the following code starting at address xxxx.
```

Example Program		Address	Hex
.org 0400h	becomes:	0400	79
MOV r2,#00h		0401	00

The .org pseudo op lets you put code and data anywhere in program memory you wish. Normally the program starts at 0000h using a .org 0000h.

```
.equ label,xxxx    EQUate the label name to the number xxxx.
```

Example Program		Address	Hex
.org 0000h	becomes:	0000	74
.equ fred,12h		0001	12
mov a,#fred			

.equ turns numbers into names; it makes the program much more readable because the name chosen for the label can have some meaning in the program, whereas the number will not.

```
.db xx    Define a Byte: place the 8-bit number xx next in memory.
```

Example Program		Address	Hex
.org 0100h	becomes:	0100	34
.db 34h		0101	56
.db 56h			

```
.db ''abc''
```

Example Program		Address	Hex
.org 0200h	becomes:	0200	31
.db "123 "		0201	32
		0202	33
		0203	20

.db xx takes the number xx (from 0 to 255d) and converts it to hex in the next memory location. .db "abc" will convert *any* character that can be typed into the space between the quote marks into the equivalent ASCII (no parity) hex code for that character, and place them sequentially in memory. .db permits the programmer to place any hex byte anywhere in memory.

```
.dw xxxx    Define a Word. Place the 16-bit number xxxx in memory.
```

Example Program		Address	Hex
.org 0abcdh	becomes:	ABCD	12
.dw 1234h		ABCE	34

.dw is a 16 bit version of .db.

```
.end    The End. Tells the assembler to stop assembling.
```

Other directives exist that are rarely used by student programmers. Refer to the assembler documentation contained in the disk file under the name LEVELI.DOC. The file INTEL1.ASM contains some .opdef directives which let .anything become anything (no .) for those programs written with directives which do not use the period.

Numbers

Numbers follow one simple rule: They must start with a number from 0 to 9. For example,

```
1234
0abcdh
0ffh
5aceh
```

Numbers in the program can be written in decimal or hex form as

```
1234 = 1234 decimal
h'0dd = DD hexadecimal
0ddh = DD hexadecimal
```

The first form of the hexadecimal number (h'0dd) is a Unix standard, while the second form (0ddh) is a common assembly language standard.

Labels

Labels are names invented by the programmer that stand for a number in the program, such as a constant in the .equ directive above, or a number which represents a memory location in the program. Labels used for memory locations follow two simple rules:

1. All labels must START with an alphabetic character and END with : (colon).

2. No more than 8 characters.

The following are examples:

```
fred:
m1:
p1234:
xyz:
```

The restriction that all numbers begin with a number is now apparent; hexadecimal numbers beginning with A to F would be mistaken by the assembler as a label and chaos would result.

$\triangleright$ COMMENTS

Anything that follows a semicolon (;) in a line of a program is ignored by the assembler. Comments *must* start with a ;. For example,

```
;this is a comment and will be ignored by the assembler
```

If you are assembling a program and get a LOT of syntax errors, you probably forgot to include a semicolon in your comments.

Typing a Line

To make the program readable, it is recommended that you type all opcodes about 10 spaces or so to the right of the left margin of your text. Start all labels at the left margin of text, and place any comments to the right of the opcode entry. The finished line should appear as follows:

```
label:    opcode    ;comment
```

Inspection of the programs included with the text will provide many clues as to what syntax is acceptable to the assembler. Experiment with the assembler by writing short programs to get a clear understanding of what each output file contains.

Symbols

David Akey has very thoughtfully included a complete symbol table for the assembler that lets the programmer use symbolic names for the 8051 Special Function Registers and individual register bits. These are called "reserved" symbols; so do *not* use any of these symbols for a label, or you will get an error message in the .LST file. Forgetting to type -s when you invoke A51.EXE will get you this table and all of your labels—at the end of your .LST file.

SYMBOL	ADDR.	SYMBOL	ADDR.	SYMBOL	ADDR.
AC	=00D6	P0.7	=0087	SM2	=009D
ACC	=00E0	P1	=0090	SP	=0081
ACC.0	=00E0	P1.0	=0090	T2CON	=00C8
ACC.1	=00E1	P1.1	=0091	T2CON.0	=00C8
ACC.2	=00E2	P1.2	=0092	T2CON.1	=00C9
ACC.3	=00E3	P1.3	=0093	T2CON.2	=00CA
ACC.4	=00E4	P1.4	=0094	T2CON.3	=00CB
ACC.5	=00E5	P1.5	=0095	T2CON.4	=00CC
ACC.6	=00E6	P1.6	=0096	T2CON.5	=00CD
ACC.7	=00E7	P1.7	=0097	T2CON.6	=00CE
B	=00F0	P2	=00A0	T2CON.7	=00CF
B.0	=00F0	P2.0	=00A0	TB8	=009B
B.1	=00F1	P2.1	=00A1	TCLK	=00CC
B.2	=00F2	P2.2	=00A2	TCON	=0088
B.3	=00F3	P2.3	=00A3	TCON.0	=0088
B.4	=00F4	P2.4	=00A4	TCON.1	=0089
B.5	=00F5	P2.5	=00A5	TCON.2	=008A
B.6	=00F6	P2.6	=00A6	TCON.3	=008B
B.7	=00F7	P2.7	=00A7	TCON.4	=008C
CPRL2	=00C8	P3	=00B0	TCON.5	=008D
CT2	=00C9	P3.0	=00B0	TCON.6	=008E
CY	=00D7	P3.1	=00B1	TCON.7	=008F
DPH	=0083	P3.2	=00B2	TF0	=008D
DPL	=0082	P3.3	=00B3	TF1	=008F
EA	=00AF	P3.4	=00B4	TF2	=00CF
ES	=00AC	P3.5	=00B5	TH0	=008C
ET0	=00A9	P3.6	=00B6	TH1	=008D
ET1	=00AB	P3.7	=00B7	TH2	=00CD
ET2	=00AD	PCON	=0087	TI	=0099
EX0	=00A8	PS	=00BC	TL0	=008A
EX1	=00AA	PSW	=00D0	TL1	=008B
EXEN2	=00CB	PSW.0	=00D0	TL2	=00CC
EXF2	=00CE	PSW.1	=00D1	TMOD	=0089
F0	=00D5	PSW.2	=00D2	TR0	=008C

Continued

SYMBOL	ADDR.	SYMBOL	ADDR.	SYMBOL	ADDR.
IE	=00A8	PSW.3	=00D3	TR1	=008E
IE.0	=00A8	PSW.4	=00D4	TR2	=00CA
IE.1	=00A9	PSW.5	=00D5	TXD	=00B1
IE.2	=00AA	PSW.6	=00D6		
IE.3	=00AB	PSW.7	=00D7		
IE.4	=00AC	PT0	=00B9		
IE.5	=00AD	PT1	=00BB		
IE.7	=00AF	PT2	=00BD		
IE0	=0089	PX0	=00B8		
IE1	=008B	PX1	=00BA		
INT0	=00B2	RB8	=009A		
INT1	=00B3	RCAP2H	=00CB		
'IP	=00B8	RCAP2L	=00CA		
IP.0	=00B8	RCLK	=00CD		
IP.1	=00B9	REN	=009C		
IP.2	=00BA	RI	=0098		
IP.3	=00BB	RS0	=00D3		
IP.4	=00BC	RS1	=00D4		
IP.5	=00BD	RXD	=00B0		
IT0	=0088	SBUF	=0099		
IT1	=008A	SCON	=0098		
OV	=00D2	SCON.0	=0098		
P	=00D0	SCON.1	=0099		
P0	=0080	SCON.2	=009A		
P0.0	=0080	SCON.3	=009B		
P0.1	=0081	SCON.4	=009C		
P0.2	=0082	SCON.5	=009D		
P0.3	=0083	SCON.6	=009E		
P0.4	=0084	SCON.7	=009F		
P0.5	=0085	SM0	=009F		
P0.6	=0086	SM1	=009E		

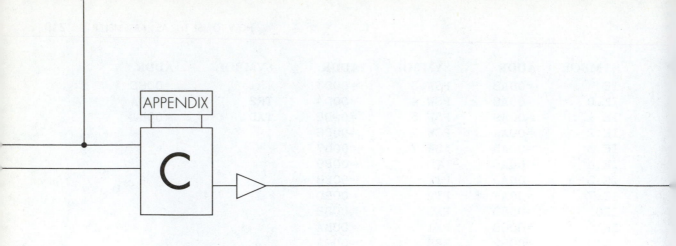

APPENDIX

C

How to Use the Simulator

Introduction

One learns to program by writing *and* testing programs. There are many ways to test a program; the most traditional is to load the program into a hardware specific target system that uses the computer under study and execute the program. Loading can range from transferring the object file from the development system computer to the target system RAM for execution via a serial data link, or programming an EPROM with the object file and inserting the EPROM into a target system memory socket for testing.

Both of these approaches use some sort of monitor program that is found in ROM in the target system. The monitors are usually adequate for simple program tests:

Instruction single step or run

Display register and memory contents

Stop at selected program addresses

This capability allows programs to be debugged in a methodical way but requires considerable skill and time by the programmer.

With the advent of affordable personal computers, programs have appeared that use personal computer resources to *simulate* the operation of the target computer; the programmer now has a unique view of the computer registers and memory *as the program runs*.

Simulators usually show the internal register and memory locations on the screen of the personal computer and allow the programmer to perform all of the operations listed above for a monitor with the added advantage of watching the data change as the program operates. This saves considerable time because the register and memory contents do not have to be displayed using separate monitor commands. The visual representation also gives the programmer a better "feel" for what is taking place in the program.

The program under test can also be loaded quickly from the same file that contains the simulator, assembler, and editor program. If the personal computer has sufficient RAM to enable RAM disk structures to be created, the process of edit, assemble, and simulate can be done in a very timely manner. Finally, the expense of buying special target systems is avoided enabling the user to "try" many different computers at minimal cost.

Simulators do not, however, generally have the ability to perform actual I/O or internal hardware operations such as timing or data transmission and reception. You must, at last, try the program in a target system when doing an actual application.

The 8051 simulator used with this book is the PseudoMax 51 written by David Akey of PseudoCode, Newport News, Virginia. David also supplied the PseudoSam 51 A51 assembler, which has been used to write all of the programs in the text and is included on the book disk.

Computer Configuration Needed to Run the Simulator

The 8051 Simulator runs on IBM PC and compatible computers. Requirements for the PC are

512K RAM

DOS version 2.1 or newer

IBM Mono, CGA, EGA, or compatible monitor

Two disk drives (One must be 5 1/4")

Features

The outstanding feature of the simulator is the ability of the user to construct screens that show various parts of an 8051 system. Each screen is made up of separate windows that display internal CPU registers and code and data memory areas. The screen set can be saved as a disk file and used for one type of problem; another screen set can be configured for a different type of problem and loaded when needed.

The user may construct up to ten screens, each made up of a mixture of the 42 available register and memory windows. Not all 42 windows can fit on one screen, so different screens must be used to show the total set of windows needed for a particular program simulation.

To run a simulation, the screen set file is loaded into the simulator first, followed by a program in object code format. The program is then run using these simulator commands:

1. Reset the program counter to 0000h
2. Single step the program
3. Free run the program
4. Free run until breakpoint is reached
5. Stop free run

The contents of any location in code ROM and internal RAM (including the special-function registers) and external memory may be changed by the user while the program runs. Port I/O may be simulated by changing the value of the port special function registers. Interrupts are simulated by striking function keys on the PC keyboard.

The simulator included with the text is a student version that is identical to the professional version with the exception that memory is limited to 3FFh bytes each of code and data address space. A professional version, which has the full memory address capability, can be obtained from PseudoCode or other authors of simulator programs.

The Simulator Programs

The disk contains two simulator program files: S51.EXE and BOOK.BSS. S51 is the PseudoMax 8051 simulator, and BOOK.BSS is a sample simulator file that contains four screens. BOOK.BSS has been used to simulate all of the programs in the text using the professional simulator version. BOOK.BSS may be used, as is, to simulate programs written in response to problems in the text; the student is encouraged to create other .BSS files once operation of the simulator has been mastered.

Starting the Simulator

Before using the simulator you should have an object file ready to simulate. Write a small program (5 or 6 lines) assemble it, and you will have a .OBJ program for trial use in the simulator. You may name it what you will; it will be identified as yourfile.obj in the instructions that follow.

After booting up your DOS system, place the disk containing the simulator in drive B, and your disk with the .OBJ file in drive A. The simulator must be loaded with two files: the screen file, Book.bss, and your object file, yourfile.obj, before it is run:

1. Go to the B> prompt and type: s51 <return>
   The program will load, and display the menu screen shown in Figure C.1. The status line will ask you to *Select*.

2. Type L for Load
   The status line will ask you to *Select an Object file or Previous machine*.

3. Type P for Previous machine.
   The status line will ask for the saved *filename*.

4. Type book.bss <return>
   book.bss will be loaded into the simulator, and the status line will ask you to *Select*.

5. Type L for Load
   The status line will ask you to *Select an Object file or Previous machine*.

6. Type O for Object file
   The status line will ask for the saved *filename*.

7. Type in a:yourfile.obj <return>
   Yourfile.obj will be loaded into the simulator, and the status line ask you to *Select*.

8. Type R for Run
   The simulator will display the first Screen, shown in Figure C.2, and await your commands.

Running the Simulation of YOURFILE.OBJ

Once the object file is loaded, you can

1. Reset the system by typing CTRL-Home.
2. Step the program by pressing the left arrow ← key. (The screen is updated after each step.)
3. Free run the program by pressing the right arrow → key. (The screen updates constantly.)
4. Speed up the free run execution time by stopping updating using function key F10. F9 restores updating.

5. Stop free run by pressing the END key or the left arrow (step) key.

6. Make changes in memory or register contents by typing commands on the screen status line.

7. Exit the Run mode, and return to the menu screen by pressing the CTRL END keys.

BOOK.BSS Simulator Screens

Four simulator screens, shown in Figures C.2 to C.5, are defined on BOOK.BSS. These screens have been designed to offer a view of many different areas of an 8031 system. A screen is chosen by pressing the alternate key and a function key simultaneously. Every screen has a status line at the bottom for typing various memory and register configuration commands.

Screen 1: (ALT–F1) The Main Screen.

The viewer can observe the operations of the PC, SP, IE, and A registers, and ports P1 and P3 in individual windows. Special-function registers DPL, DPH, PCON, TCON, TMOD, TL0, TL1, TH0, and TH1 can be found in the internal RAM window 2, which displays a portion of the SFR area. Internal RAM window 1 shows register banks 0 and 1. An instruction execution window will display program mnemonics as the program is operated.

Screen 2: (ALT–F2) The Internal RAM Screen.

Internal RAM windows 3 to 5 display internal RAM from 10h to 3Fh. Window 6 shows the SFR area, which includes SCON and SBUF. The SP, DPTR, A, and PC are also shown.

Screen 3: (ALT–F3) The ROM Screen.

Program code addresses from 0000h to 00BFh are displayed in code memory windows 1 to 3. The PC, A, DPTR, and instruction execution windows are also part of this screen.

Screen 4: (ALT–F4) The External RAM Screen.

External RAM from addresses 0000h to 00BFh are displayed in external data memory windows 1 to 3. The PC, A, DPTR, and instruction execution windows are also part of this screen.

Changing Register and Memory Contents

As the program runs you may wish to change the contents of a register or memory address:

1. Change any register contents by typing REGNAM = XX <return> on the status line (where REGNAM is one of the register names given in Appendix B.3 and XX is any hexadecimal data). For example, P1=AA will load the P1 Window with the data AAh.

2. Change External RAM contents by typing &ADDRESS = XX

3. Change Internal RAM contents by typing: *ADDRESS = XX

4. Change Code ROM contents by typing: @ADDRESS = XX

Here, ADDRESS is any legal address from 0000 to 03FF, and the following are examples:

&0040=.BC loads external address 40 with data BCh.

*01= 12 loads internal address 01 with data 12h.

(You could also type R1=12 IF Bank 0 is selected)

@00C0= B6 changes code address C0 contents to B6h.

To change entire blocks of memory, do not enter the =XX part of the line.

Setting Breakpoints

Breakpoints are memory addresses that cause the program to stop when in the free run operating mode. The program will free run until a breakpoint address is accessed in any way (read from it, write to it, or fetch it for program execution) and then stop.

The program can be started in free run again and will run until the next breakpoint address is reached. Breakpoints are typed on the status line.

1. Set a breakpoint in code memory by typing: !Address = +b
2. Set a breakpoint in internal RAM by typing: #Address = +b
3. Set a breakpoint in external memory by typing: %Address = +b
4. Disable a breakpoint in memory by typing: (!,#,%)Address = −b

Here, Address = 0000 to 03FF. For example, !0040 = +b will set a breakpoint at program address 40, while !0040 = −b will clear it. Note that I/O ports are internal RAM and can set with the breakpoint attribute. I/O operations automatically stop the Free Run.

Generating Interrupts

All of the 8051 interrupts can be simulated by using the function keys while pressing the shift key:

```
IE0 = SHIFT F1
TF0 = SHIFT F2
IE1 = SHIFT F3
TF1 = SHIFT F4
RI,TI = SHIFT F5
```

Saving a Session

To save the present state of a session, exit the run mode by using the CTRL END keys and return to the main menu. Press the S key and save your simulation state by following the prompts and naming the program yourfile.bss. When re-starting the simulator, use the name yourfile.bss instead of book.bss when loading the Previous machine file. The .obj program has been saved also, so you may proceed directly to run after loading yourfile.bss.

Creating Your Own Screens

To create custom screens you should invoke s51 and load BOOK.BSS. Then, when prompted by the status line, press P for "Profile." Profile is the process of creating your own screens. While in the profile mode the status line for each screen will display your choices: Add, Delete, Move, Copy, Quit.

The first Screen of BOOK.BSS will appear, and you can begin to configure your first screen by (D)eleting windows and (A)dding windows, then (M)oving them around. Use the ALT−FX keys to go to screens 1 to 10.

Windows are deleted by positioning the cursor box on an undesired window (using the right and left arrow keys) and pressing the D key. Windows are added by pressing the A key and typing in a window number at the prompt on the status line.

The numbers for each Window are as follows:

Window	Number	Window	Number
A register	16	Program counter	17
B register	18	Stack pointer	21
Data pointer	22	Port 0	23
Port 1	24	Port 2	25
Port 3	26	IP register	27
IE register	28	PCON register	29
Stack	30	External RAM 1–8	31–38
Code ROM 1–8	39–46	Internal RAM 1–8	47–54
PSW register	55	SCON register	57
TCON	58	Execution	59

For example, screen 1 of BOOK.BSS consists of windows 16, 17, 59, 28, 55, 24, 26, 30, 47, and 48.

The window can be moved anywhere on the screen by pressing the M key and using the cursor keys (up, down, left, right) to position the window at the desired screen location. A return key fixes the new location chosen.

When a screen is done, go to the next screen and repeat the same steps. Pressing the Q key will return to the original s51 menu.

The copy command will copy the screen number typed in response to the status line query to the current screen. Normally the current screen will be empty and a previously done screen copied to it.

Setting RAM and ROM Window Starting Addresses

After making your screens you may also select the beginning address of each memory window. An inspection of BOOK.BSS internal RAM window 2 of screen 1 shows that the beginning address is 82h, which is the internal address of the DPL special-function register.

To set a memory beginning address you must be in the (R)un mode. Select a screen with the memory window of interest and type on the status line:

```
.mw# = xxxx    .irw# = xx    .edw# = xxxx
```

Here, # = 1 to 8, .mw is code memory window, .irw is internal RAM window, and .edw is external data window. For example, to set Internal Ram Window 2 to start at address 82h on screen 1 of BOOK.BSS, the command typed on the status line of screen 1 when in the run mode is:

```
.irw2 = 82 <return>
```

Setting Memory Attributes

The last task to be performed is to determine the type of memory access for each byte of memory. Memory can have these attributes:

r	read
w	write
e	execute
io	input/output

b   breakpoint
n   ignore the rest (memory-type dependent)

An attribute can be enabled by typing a + (plus) in front of it. The section on setting breakpoints shows any memory address can cause the program to stop in free run by a +b. Attributes are disabled by a − (minus) in front of the attribute letter; −b removes a break-point from an address.

BOOK.BSS has the following attributes assigned:

Code memory	+r −w +e −io −b −n
	(+n = +r +w +e −io −b)
Internal RAM	+r +w +io −b −n
	(+n = +r +w +io −b)
External data	+r +w −e −io −b −n
	(+n = +r +w +e −io −b)

To view these attributes (you must be in the run mode), type ! for code, # for internal RAM, or % for external data followed by a return. A window will pop up on the screen showing the memory addresses and attributes. The page up and page down keys can scan lengthy memory attribute windows.

To change an attribute of one address or a range of addresses, type:

```
!start address..end address = +/- rweiobn for the Code addresses
#start address..end address = +/- rwiobn for the Internal Ram
%Start address..end address = +/- rweiobn for External Data
```

For example, BOOK.BSS memory attributes were set by typing:

```
!0000..03ff = +r -w +e -io -b -n for Code Memory
#00..ff = +r +w -io -b -n for Internal Ram
#80..80 = +io for Port 0 of internal Ram
#90..90 = +io for Port 1 of internal Ram
#A0..A0 = +io for Port 2 of internal Ram
#B0..B0 = +io for Port 3 of internal Ram
%0000..03ff = +r +w -e -io -b -n for the External Data
```

Forgetting to set the attributes correctly (such as making the code memory −e) will cause the simulator to stop while in free run and display a violation in the upper right hand corner of the screen.

Remember:

After having configured your screens, set the memory starting addresses, and given attributes to all memory, save your new .BSS file by leaving the run mode (CTRL END) to get to the s51 Menu screen. Press the S key to save the new file and provide the new file name on the status line when asked.

FIGURE C.1   Menu

```
PseudoMax™ 51 ''the smart simulator''
Educational Version B1.1.01
Copyright © 1990 PseudoCorp
All rights reserved!
Licensed for non-commercial Educational use only!
```

```
DIRECTORY
RUN
SAVE
LOAD
QUIT
PROFILE
```

```
Select:
```

FIGURE C.2 Main Screen

Screen: 1 Trace: OFF   Update   STOP

Internal Ram Window 1

Addr	Value	Addr	Value
R0	= 00	0008	= 00
R1	= 00	0009	= 00
R2	= 00	000A	= 00
R3	= 00	000B	= 00
R4	= 00	000C	= 00
R5	= 00	000D	= 00
R6	= 00	000E	= 00
R7	= 00	000F	= 00

SP	Stack
05	00
06	00
07	00
08	00
09	00

Internal Ram Window 2

Addr	Value	Addr	Value
DPL	= 00	TL0	= 00
DPH	= 00	TL1	= 00
0084	= 00	TH0	= 00
0085	= 00	TH1	= 00
0086	= 00	008E	= 00
PCON	= 00	008F	= 00
TCON	= 00	P1	= FF
TMOD	= 00	0091	= 00

P1	
FF	
FF	
FF	
FF	

P3	
FF	
FF	
FF	
FF	

PC	A
0000	00
0000	00
0000	00
0000	00
0000	00

Loc	Inst Addr
0000	
0000	
0000	
0000	
0000	

IE	
00	
00	
00	
00	
00	

CY	AC	F0	RS1	RS0	OV	–	P
0	0	0	0	0	0	0	0
0	0	0	0	0	0	0	0
0	0	0	0	0	0	0	0
0	0	0	0	0	0	0	0
0	0	0	0	0	0	0	1

FIGURE C.3 Main Screen 1

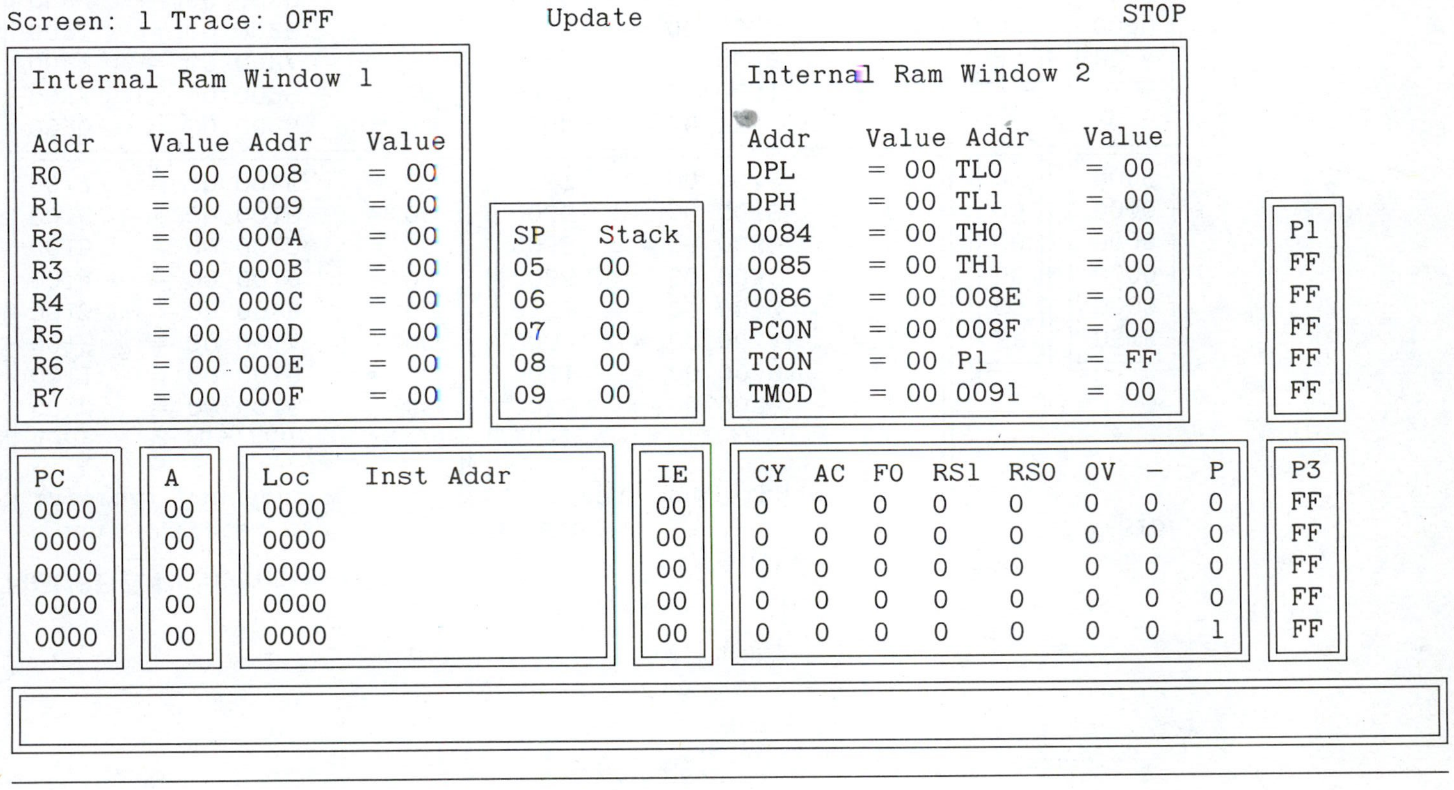

Screen: 1 Trace: OFF   Update   STOP

```
Internal Ram Window 1

Addr   Value  Addr   Value
R0   = 00     0008 = 00
R1   = 00     0009 = 00
R2   = 00     000A = 00
R3   = 00     000B = 00
R4   = 00     000C = 00
R5   = 00     000D = 00
R6   = 00     000E = 00
R7   = 00     000F = 00
```

```
Internal Ram Window 2

Addr   Value  Addr   Value
DPL  = 00     TL0  = 00
DPH  = 00     TL1  = 00
0084 = 00     TH0  = 00
0085 = 00     TH1  = 00
0086 = 00     008E = 00
PCON = 00     008F = 00
TCON = 00     P1   = FF
TMOD = 00     0091 = 00
```

```
SP  Stack
05  00
06  00
07  00
08  00
09  00
```

```
P1
FF
FF
FF
FF
FF
```

```
P3
FF
FF
FF
FF
FF
```

CY	AC	FO	RS1	RS0	OV	-	P
0	0	0	0	0	0	0	0
0	0	0	0	0	0	0	0
0	0	0	0	0	0	0	0
0	0	0	0	0	0	0	1

```
IE
00
00
00
00
00
```

```
PC     A
0000   00
0000   00
0000   00
0000   00
0000   00
```

```
Loc    Inst Addr
0000
0000
0000
0000
0000
```

FIGURE C.4 Internal RAM Screen 2

```
Screen: 2 Trace: OFF            Update                                  STOP
 Internal Ram Window 3          Internal Ram Window 5

 Addr    Value Addr    Value     Addr    Value Addr    Value
 0010    = 00  0018    = 00      0030    = 00  0038    = 00
 0011    = 00  0019    = 00      0031    = 00  0039    = 00      SP     DPTR
 0012    = 00  001A    = 00      0032    = 00  003A    = 00      00     0000
 0013    = 00  001B    = 00      0033    = 00  003B    = 00      00     0000
 0014    = 00  001C    = 00      0034    = 00  003C    = 00      00     0000
 0015    = 00  001D    = 00      0035    = 00  003D    = 00      00     0000
 0016    = 00  001E    = 00      0036    = 00  003E    = 00      07     0000
 0017    = 00  001F    = 00      0037    = 00  003F    = 00

 0020    = 00  0028    = 00      SCON    = 00  P2      = FF
 0021    = 00  0029    = 00      SBUF    = 00  00A1    = 00
 0022    = 00  002A    = 00      009A    = 00  00A2    = 00      A      PC
 0023    = 00  002B    = 00      009B    = 00  00A3    = 00      00     0000
 0024    = 00  002C    = 00      009C    = 00  00A4    = 00      00     0000
 0025    = 00  002D    = 00      009D    = 00  00A5    = 00      00     0000
 0026    = 00  002E    = 00      009E    = 00  00A6    = 00      00     0000
 0027    = 00  002F    = 00      009F    = 00  00A7    = 00      00     0000
```

FIGURE C.5 Code Screen 3

Screen: 3 Trace: OFF

Update   STOP

Code Memory Window 1

	0	1	2	3	4	5	6	7
	8	9	A	B	C	D	E	F
0000	00	00	00	00	00	00	00	00
	00	00	00	00	00	00	00	00
0010	00	00	00	00	00	00	00	00
	00	00	00	00	00	00	00	00
0020	00	00	00	00	00	00	00	00
	00	00	00	00	00	00	00	00
0030	00	00	00	00	00	00	00	00
	00	00	00	00	00	00	00	00

Code Memory Window 2

	0	1	2	3	4	5	6	7
	8	9	A	B	C	D	E	F
0040	00	00	00	00	00	00	00	00
	00	00	00	00	00	00	00	00
0050	00	00	00	00	00	00	00	00
	00	00	00	00	00	00	00	00
0060	00	00	00	00	00	00	00	00
	00	00	00	00	00	00	00	00
0070	00	00	00	00	00	00	00	00
	00	00	00	00	00	00	00	00
0080	00	00	00	00	00	00	00	00
	00	00	00	00	00	00	00	00
0090	00	00	00	00	00	00	00	00
	00	00	00	00	00	00	00	00
00A0	00	00	00	00	00	00	00	00
	00	00	00	00	00	00	00	00
00B0	00	00	00	00	00	00	00	00
	00	00	00	00	00	00	00	00

PC	A	Loc	Inst Addr	DPTR
0000	00	0000		0000
0000	00	0000		0000
0000	00	0000		0000
0000	00	0000		0000
0000	00	0000		0000

FIGURE C.6 External RAM Screen 4

Screen: 4 Trace: OFF   Update   STOP

```
Extern Data Memory Window 1
      0 1 2 3 4 5 6 7
      8 9 A B C D E F
0000 00 00 00 00 00 00 00 00
     00 00 00 00 00 00 00 00
0010 00 00 00 00 00 00 00 00
     00 00 00 00 00 00 00 00
0020 00 00 00 00 00 00 00 00
     00 00 00 00 00 00 00 00
0030 00 00 00 00 00 00 00 00
     00 00 00 00 00 00 00 00
```

```
Extern Data Memory Window 2
      0 1 2 3 4 5 6 7
      8 9 A B C D E F
0040 00 00 00 00 00 00 00 00
     00 00 00 00 00 00 00 00
0050 00 00 00 00 00 00 00 00
     00 00 00 00 00 00 00 00
0060 00 00 00 00 00 00 00 00
     00 00 00 00 00 00 00 00
0070 00 00 00 00 00 00 00 00
     00 00 00 00 00 00 00 00

0080 00 00 00 00 00 00 00 00
     00 00 00 00 00 00 00 00
0090 00 00 00 00 00 00 00 00
     00 00 00 00 00 00 00 00
00A0 00 00 00 00 00 00 00 00
     00 00 00 00 00 00 00 00
00B0 00 00 00 00 00 00 00 00
     00 00 00 00 00 00 00 00
```

```
PC      A       Loc   Inst  Addr      DPTR
0000    00      0000                   0000
0000    00      0000                   0000
0000    00      0000                   0000
0000    00      0000                   0000
0000    00      0000                   0000
```

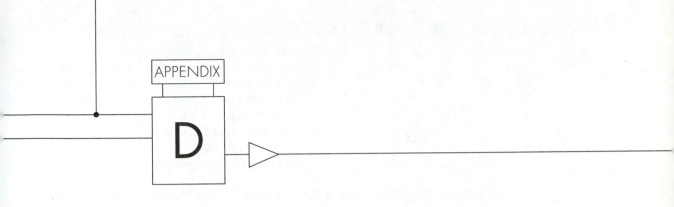

The 8255 Programmable I/O Port

Introduction

Eight-bit microprocessor families included peripheral chips that are used with the CPU to provide many of the I/O functions that are now found integrated inside a microcontroller. As the 8-bit microprocessor fades into obsolescence, these peripheral chips are finding a new life in augmenting microcontroller I/O capability.

These peripheral chips include serial and parallel I/O as well as interrupt controllers and dynamic RAM controllers. The 8051 loses two parallel I/O ports when used with external memory, and part of a third to serial data communication and interrupt functions. To make up for this loss, a programmable parallel port chip, the 8255, is often added to an 8051 system, as discussed in Chapter 7. This appendix describes how to use the 8255 as a basic parallel I/O port. The 8255 is capable of many sophisticated I/O functions, including interrupts and handshaking. Refer to the manufacturers' literature for a complete description of 8255 capabilities and programming.

Functional Description

The 8255 features three 8-bit programmable parallel I/O ports named A, B, and C. Port C can be used as two separate ports of four bits each if properly programmed.

Before any port can be used, the 8255 must be programmed by writing the proper control bits to the control register. The three ports may then be accessed by the 8051 program. The 8255 uses the address lines A0 and A1 to access the Control register and the three Ports. The $\overline{RD}$, $\overline{WR}$ and $\overline{CS}$ lines are enabled by the particular decoding scheme used by the 8051 system designer. The resulting control and address states yield the following actions:

A1	A0	$\overline{RD}$	$\overline{WR}$	$\overline{CS}$	Action
0	0	L	H	L	Read the Contents of port A
0	1	L	H	L	Read the Contents of port B
1	0	L	H	L	Read the Contents of port C
0	0	H	L	L	Write to the port A Latch
0	1	H	L	L	Write to the port B Latch
1	0	H	L	L	Write to the port C Latch
1	1	H	L	L	Write to the Control register
X	X	X	X	H	Data bus to high impedance

The 8255 appears much like an internal port of the 8051 once it has been programmed.

Programming The 8255

Control bytes written to the control register use each bit of the byte to program some feature of the 8255:

Bit	State	Result
7	1	Program ports for mode and input or output
7	0	Set/reset individual bits of port C

When bit 7 is a 1 then the ports are programmed as:

Bit	State	Result
6,5	00	Set port A and C4–C7 in I/O mode 0
6,5	01	Set port A and C4–C7 in I/O mode 1
6,5	10	Set port A and C4–C7 in I/O mode 2
4	0	Set port A as an output port
4	1	Set port A as an input port
3	0	Set C4–C7 as an output port
3	1	Set C4–C7 as an input port
2	0	Set port B and C0–C3 in I/O mode 0
2	1	Set port B and C0–C3 in I/O mode 1
1	0	Set port B as an output port
1	1	Set port B as an input port
0	0	Set C0–C3 as an output port
0	1	Set C0–C3 as an input port

8255 I/O Modes

Port A and the high part of port C may be programmed in one of three modes, port B and the lower part of port C may be programmed in one of two modes. The modes are

Mode 0—Basic I/O: Data written to the port is latched; data read from the port is read from the input pins. (This mode is identical to 8051 port operation.)

Mode 1—Strobed I/O: This handshaking mode uses ports A and B as I/O and port C to generate handshaking signals to the devices connected to ports A and B and an interrupt signal to the host microcontroller.

Mode 2—Strobed bi-directional I/O: This mode is similar to Mode 1 with the ability to use port A as a bi-directional data bus.

Modes 1 and 2 require setting interrupt enable bits in the port C data register. These modes are intended to be used with intelligent peripherals such as printers.

Reset Condition

Upon reset all the port data latches and the control register contents are cleared to 00. The ports are all in the input mode.

Control Registers

Introduction

For the convenience of the programmer the control special-function register figures from Chapter 2 are shown here for easy reference. An ASCII table is shown on the following page.

TCON and TMOD Function Registers

7	6	5	4	3	2	1	0
TF1	TR1	TF0	TR0	IE1	IT1	IE0	IT0

7	6	5	4	3	2	1	0
Gate	C/$\overline{T}$	M1	M0	Gate	C/$\overline{T}$	M1	M0

[Timer 1] [Timer 0]

SCON and PCON Function Registers

7	6	5	4	3	2	1	0
SM0	SM1	SM2	REN	TB8	RB8	TI	RI

7	6	5	4	3	2	1	0
SMOD	—	—	—	GF1	GF0	PD	IDL

IE and IP Function Registers

7	6	5	4	3	2	1	0
EA	—	ET2	ES	ET1	EX1	ET0	EX0

7	6	5	4	3	2	1	0
—	—	PT2	PS	PT1	PX1	PT0	PX0

ASCII Codes for Text and Control Characters—No Parity

HEX	Character	HEX	Character	HEX	Character	HEX	Character
00	NUL	28	(	50	P	78	x
01	SOH	29	)	51	Q	79	y
02	STX	2A	*	52	R	7A	z
03	ETX	2B	+	53	S	7B	{
04	EOT	2C	,	54	T	7C	|
05	ENQ	2D	–	55	U	7D	}
06	ACK	2E	.	56	V	7E	~
07	BEL	2F	/	57	W	7F	(del)
08	BS	30	0	58	X		
09	HT	31	1	59	Y		
0A	LF	32	2	5A	Z		
0B	VT	33	3	5B	[		
0C	FF	34	4	5C	\		
0D	CR	35	5	5D	]		
0E	SO	36	6	5E	^		
0F	SI	37	7	5F	—		
10	DLE	38	8	60	`		
11	DC1	39	9	61	a		
12	DC2	3A	:	62	b		
13	DC3	3B	;	63	c		
14	DC4	3C	<	64	d		
15	NAK	3D	=	65	e		
16	SYN	3E	>	66	f		
17	ETB	3F	?	67	g		
18	CAN	40	@	68	h		
19	EM	41	A	69	i		
1A	SUB	42	B	6A	j		
1B	ESC	43	C	6B	k		
1C	FS	44	D	6C	l		
1D	GS	45	E	6D	m		
1E	RS	46	F	6E	n		
1F	US	47	G	6F	o		
20	(space)	48	H	70	p		
21	!	49	I	71	q		
22	"	4A	J	72	r		
23	#	4B	K	73	s		
24	$	4C	L	74	t		
25	%	4D	M	75	u		
26	&	4E	N	76	v		
27	'	4F	O	77	w		

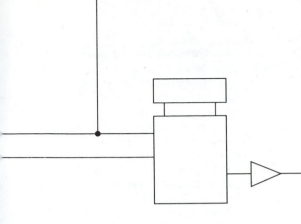

Index

239

**IMPORTANT: PLEASE READ BEFORE OPENING DISKETTE PACKAGE.
THIS TEXT IS NOT RETURNABLE IF SEAL IS BROKEN.**

SIMULATOR PROGRAM
LIMITED USE LICENSE

By accepting this license, you have the right to use the Software and the accompanying documentation, but you do not become the owner of these materials.

1. PERMITTED USES

You are granted a non-exclusive limited license to use the Software under the terms and conditions stated in this License. You may:

    a.   Use the Software on a single computer.
    b.   Make a single copy of the Software in machine readable form solely for backup purposes in support of your use of the Software on a single machine. You must reproduce and include the copyright notice on any copy you make.
    c.   Transfer this copy of the Software and the License to another user if the other user agrees to accept the terms and conditions of this License. If you transfer this copy of the Software you must also destroy the backup copy you made. Transfer of this copy of the Software and the License automatically terminates this License as to you.

2. PROHIBITED USES

You may not use, copy, modify, distribute or transfer the Software or any copy, in whole or in part, except as expressly permitted in this License.

3. TERMS

This License is effective when you open the diskette package and remains in effect until terminated. You may terminate this License at any time by ceasing all use of the Software and destroying this copy and any copy you have made. It will also terminate automatically if you fail to comply with the terms of this License. Upon termination, you agree to cease all use of the Software and destroy all copies.

4. DISCLAIMER OF WARRANTY

Except as stated herein, the Software is licensed "as is" without warranty of any kind, express or implied, including warranties of merchantability or fitness for a particular purpose. You assume the entire risk as to the quality and performance of the Software. You are responsible for the selection of the Software to achieve your intended results and for the installation, use, and results obtained from it. West Publishing and West Services do not warrant the performance of nor results that may be obtained with the Software. West Services does warrant that the diskette(s) upon which the Software is provided will be free from defects in materials and workmanship under normal use for a period of thirty days from the date of delivery to you as evidenced by a receipt. Some states do not allow the exclusion of implied warranties so the above exclusion may not apply to you. This warranty gives you specific legal rights. You may also have other rights which vary from state to state.

5. LIMITATION OF LIABILITY

Your exclusive remedy for breach by West Services of its limited warranty shall be replacement of any defective diskette upon its return to West at the above address, together with a copy of the receipt, within the warranty period. If West Services is unable to provide you with a replacement diskette which is free of defects in material and workmanship, you may terminate this License by returning the Software and the license fee paid hereunder will be refunded to you. In no event will West be liable for any lost profits or other damages including direct, indirect, incidental, special, consequential or any other type of damages arising out of the use or inability to use the Software even if West Services has been advised of the possibility of such damages.

6. GOVERNING LAW

This Agreement will be governed by the laws of the State of Minnesota.

You acknowledge that you have read this License and agree to its terms and conditions. You also agree that this License is the entire and exclusive agreement between you and West and supersedes any prior understanding or agreement, oral or written, relating to the subject matter of this Agreement.

West Publishing Company